Frank Rutley

The Study of Rocks

An Elementary Textbook on Petrology. Second Edition

Frank Rutley

The Study of Rocks
An Elementary Textbook on Petrology. Second Edition

ISBN/EAN: 9783337276485

Printed in Europe, USA, Canada, Australia, Japan

Cover: Foto ©berggeist007 / pixelio.de

More available books at **www.hansebooks.com**

THE STUDY OF ROCKS

AN ELEMENTARY TEXT-BOOK ON PETROLOGY

By FRANK RUTLEY, F.G.S.

H. M. GEOLOGICAL SURVEY

SECOND EDITION

D. APPLETON AND CO.
NEW YORK
1881

PREFACE

TO

THE SECOND EDITION.

IN THIS EDITION, such corrections and additions as seemed necessary have been inserted. But little alteration has been made in the chapter on the optical properties of minerals, since the subject will be more fully expounded in the Text-Books of Light and Mineralogy which will shortly be published in this series. A table of the extinctions in the Felspars is added at p. 310, Appendix B.

F. R.

February 1880.

PREFACE
TO
THE FIRST EDITION.

THE rapid advance of Petrological study during the last few years has rendered it imperative that some English text-book should be written for the guidance of students in this branch of science. Several good manuals of petrology have recently been published on the Continent; but, hitherto, comparatively little has been done, in this country, to supply elementary instruction in the systematic study of rocks. The application of the microscope, in this special branch of geology, has of late years afforded more precise information, concerning the mineral constitution and minute structure of rocks, than it was possible to acquire by the older methods of research; and, in this book, I have endeavoured to give a clear explanation of the method of preparing sections of rock for microscopic examination, as well as a description of the microscopic characters of the most important rock-forming minerals, upon the identification of which the determination of the precise character of a rock is necessarily based. I have been compelled to make very free use of foreign

Preface to the First Edition.

works on petrology, especially those of Zirkel, Rosenbusch, and Von Lasaulx, and I have also extracted much information from other foreign and British publications. In all instances I have endeavoured to indicate the sources from which the information has been derived, and in this respect I trust that no injustice has been done to any author. I am also greatly indebted to Professor A. Rénard, of the Royal Museum in Brussels, for the revision of one or two chapters and for many useful hints. I would especially thank Professor John Morris for his kindness in voluntarily undertaking the revision of the entire work; indeed I cannot adequately express my obligation to him for his friendly and valuable criticisms.

Professor Zirkel has kindly given me permission to copy some of the microscopic drawings from his works, and has also assisted me by sending me some of his publications.

To Professors J. W. Judd, A. H. Green, and T. G. Bonney, and to Messrs. H. W. Bristow, J. A. Phillips, H. Bauerman, S. Allport, W. Chandler Roberts, Trenham Reeks, E. Best, T. V. Holmes, and others, I am also indebted for information or for specimens which have helped me in the prosecution of my work, while from the Editor, Mr. C. W. Merrifield, I have received a multitude of useful suggestions.

My thanks are likewise due to Mr. E. T. Newton,

Assistant Naturalist to the Geological Survey, for some interesting notes on his method of preparing microscopic sections of coal, and to Mr. J. B. Jordan, of the Mining Record Office, for the revision of the chapter on the preparation of microscopic sections, and for the use of the wood-block representing the construction of his section-cutting machine.

In the classification I have to some extent deviated from the systems commonly adopted; and, in the general treatment of the different subjects, original ideas and observations are more or less plentifully interwoven with the information derived from books.

From the limited size of this work I have necessarily been compelled to treat certain portions of the subject with brevity, but I trust that nothing of importance to the student has been omitted.

I have intentionally avoided anything more than casual references to localities where particular rocks occur, and have preferred to devote additional space to the descriptions of the typical rocks themselves.

If I appear to have entered too much into microscopical details, I can merely observe that nature makes no difference between great and small; that the great features which diversify the earth's surface, and which appear stupendous to our finite perceptions, are absurdly trivial when compared with the dimensions of the globe itself, while the latter, in relation to the sun, is a mere speck. Minute structure and gross are alike governed in their development by the same natural forces, which are giants commanding legions of atoms, and these hosts of pigmies constitute the world. If the present power of assisting vision were amplified thousands of times, we should probably find similarly perfect results, governed by the same laws, the general principle of which seems remotely hidden in those fields of inquiry, which fade away on every side into the regions separating human reason from Omniscience.

In conclusion, I may add that petrology cannot be learnt merely by reading, and that this little work does not pretend to be more than a rudimentary guide to the subject.

F. R.

CONTENTS.

PART I.

THE RUDIMENTS OF PETROLOGY.

CHAPTER I.

METHOD OF RESEARCH, ETC.

PAGE

Introductory remarks on methods of petrological research, and on the scientific and practical value of observations on the chemical and mineralogical constitution of rocks 1

CHAPTER II.

ROCKS DEFINED AND THEIR ORIGIN CONSIDERED.

Rocks are mineral aggregates—Conditions of aggregation—Preliminary considerations bearing on the origin of rocks . . 6

CHAPTER III.

DISTURBANCES OF THE EARTH'S CRUST—STRUCTURAL PLANES—SEDIMENTARY ROCKS—STRATIGRAPHY.

Subterranean forces—Vulcanicity and Seismology—Evidence of the internal heat of the earth—Fissuring and displacement of rock-masses — Structural planes — Faults — Joints — Laminar fission—Cleavage—Columnar and spheroidal structure—Rocks divided into sedimentary and eruptive groups—Stratification—

Mode of formation of sedimentary rocks—Clays, sands, limestones, &c.—Classification of sedimentary rocks—Dip and strike of beds—Flexure of beds—The influence of dip and erosion on the geological map of a country—Measurement of the thickness of beds—Denudation, marine and atmospheric—Cliffs and escarpments—Weathering of rocks dependent on relative hardness—Hills and valleys—Erosion of limestones—Weathering of eruptive rocks—Formations—Palæontological considerations—Unconformities 9

CHAPTER IV.

GENERAL CHARACTER AND MODE OF OCCURRENCE OF ERUPTIVE ROCKS.

Volcanic and plutonic rocks—Basic and acid rocks—Origin of slaty cleavage—Foliation—Metamorphism—Definition of a volcano—Volcanic phenomena 32

CHAPTER V.

THE COLLECTING AND ARRANGEMENT OF ROCK SPECIMENS.

On the collecting, dressing, labelling, and arrangement of rock specimens 39

CHAPTER VI.

PRELIMINARY EXAMINATION OF ROCKS.

Hardness tests—Description of needful apparatus, &c. . . 44

CHAPTER VII.

THE MICROSCOPE AND ITS ACCESSORIES.

Microscopes suitable for petrological work — Points essential in the construction of, and needful apparatus for such microscopes—Goniometric measurements—Stauro-microscope of Prof. Rosenbusch—Inverted microscope 46

CHAPTER VIII.

METHOD OF PREPARING SECTIONS OF MINERALS AND ROCKS FOR MICROSCOPIC EXAMINATION.

Description of needful materials and apparatus—Preparation of chips and slices—Preliminary grinding—Cementing to glass—Grinding slab—Second grinding process—Final grinding—Removal of section from grinding slab—Cleansing and mounting process—Preparation of sections of very soft rocks—Mr. E. T. Newton's method of preparing thin sections of coal . . 59

CHAPTER IX.

ON THE EXAMINATION OF THE OPTICAL CHARACTERS OF THIN SECTIONS OF MINERALS UNDER THE MICROSCOPE.

Phenomena of polarisation—Stauroscopic phenomena . . . 74

CHAPTER X.

THE PRINCIPAL ROCK-FORMING MINERALS: THEIR MEGASCOPIC AND MICROSCOPIC CHARACTERS.

Species of the felspar group—Nepheline—Leucite—Scapolite and Meionite—Sodalite, Hauyne, and Nosean—Olivine—Hypersthene—Enstatite—Bronzite—Species of the pyroxene group—Species of the amphibole group—Species of the mica group—Chlorite—Talc—Tourmaline—Epidote—Sphene—Species of the garnet group—Topaz—Zircon—Andalusite and Kyanite—Apatite—Rutile—Cassiterite—Calcspar—Quartz, &c.—Magnetite—Titaniferous iron—Hematite—Limonite—Iron and copper pyrites—Zeolites—Viridite—Opacite, &c. 86

PART II.

DESCRIPTIVE PETROLOGY.

CHAPTER XI.

THE CLASSIFICATION OF ROCKS—ERUPTIVE ROCKS, CLASS I.: VITREOUS ROCKS.

PAGE

Description of structures developed in vitreous rocks—Obsidian—Pumice—Perlite—Pitchstone—Tachylyte 174

CHAPTER XII.

ERUPTIVE ROCKS, CLASS II.: CRYSTALLINE ROCKS.

Granitic group—Syenite group—Trachyte group—Phonolite group—Andesite group—Porphyrite group—Diorite group . . 202

CHAPTER XIII.

ERUPTIVE ROCKS, CLASS II.: CRYSTALLINE ROCKS (*cont.*).

Diabase group—Gabbro group—Basalt group—Rocks of exceptional mineral constitution, including garnet rock, eklogite, lherzolite, &c.—Volcanic ejectamenta—Altered eruptive rocks. 244

CHAPTER XIV.

SEDIMENTARY ROCKS.

Unaltered series: Arenaceous, argillaceous, and calcareous groups—Altered series, including porcelain-jasper: chiastolite- and staurolite-slates, quartzite, gneiss, granulite; mica-, chlorite-, talc-, hornblende-, and other crystalline schists—Conglomerates, breccias, tufas, and sinters—Mineral deposits constituting rock-masses 274

APPENDIX 307
INDEX 315

THE STUDY OF ROCKS.

PART I.

THE RUDIMENTS OF PETROLOGY.

CHAPTER I.

METHODS OF RESEARCH, ETC.

THE means at the disposal of the older petrologists for identifying the mineral components of fine grained or minutely crystalline rocks were so primitive, that we wonder, not so much at the little that was known about them, as at the quantity of information amassed by such simple methods, and at the truth or comparative accuracy of many of their statements bearing directly upon this subject. The pocket lens was one of their most important implements in this work, and was indeed the only means they possessed for distinguishing minute structure; for although compound microscopes were known and used for physiological purposes, still the idea of slicing and grinding down fragments of rock into thin sections had not at that time occurred to anyone, or, if it had done so, had at all events never been carried into practice. Chemical analysis and simple tests of hardness, specific gravity, &c., such as have been given in treatises on mineralogy for considerably more than half a century, were the other methods which they were enabled to call in to their assistance; but chemical analysis of aggregates of minute and undetermined minerals served only to throw a very imperfect light upon the precise nature of the component minerals themselves, and in this way rocks which differed widely in minute structure and in mineral composition often yielded almost identical results so far as their ultimate chemical composition was concerned, while a knowledge of the physical conditions which governed them at the time of their deposition or solidification were matters which could only be inferred from observations made in the field, upon their mode of occurrence and relation to other rocks. Vague speculations were discussed with an energy which shows how deeply these pioneers of geology were interested in this branch of their science, but from the difficulties which attended the successful prosecution of these inquiries, especially so far as the eruptive rocks were concerned, the study of petrology as a special geological subject seemed to lapse from a state of misplaced energy into one of hopeless torpor. The first steps which in this country tended to beget fresh ardour in this direction were the publication of a paper by H. C. Sorby, in the 'Quarterly Journal of the Geological Society of London,' vol. xiv. p. 453, 'On the Microscopical Structure of Crystals, indicating the Origin of Minerals and Rocks,' and a short article by the late David Forbes, in the 'Popular Science Review,' Oct. 1867, entitled 'The Microscope in Geology.' The observations recorded in these papers were based upon the microscopic examination of thin sections of minerals and rocks; and although Mr. Sorby appears to have been the first to apply this kind of examination to purely mineralogical and petrological questions, still the method of grinding such thin sections for microscopic work was first practised by H. Witham in 1831, when conducting researches on the minute internal structure of fossil plants. The great advantage derived from the examination of thin sections of minerals lies in the circumstance, that in many cases a mineral which in ordinary hand-specimens, in thick splinters or in thick slices, would appear to be opaque, is rendered more or less translucent or transparent, the transparency increasing with the thinness of the section, so that structures which in many instances could not be discerned by reflected light are rendered apparent when the specimen is thin enough to be moderately translucent, while, in conjunction with the microscope, the polariscope, spectroscope, and goniometer may be used, and additional facilities are thus given for examining the optical properties of the mineral. Such advantages accrue from the preparation of thin slices of rocks, the component minerals of which may be studied microscopically, their crystallographic, optical, and other physical properties noted, even in rocks whose texture is so fine that examination with ordinary hand lenses is insufficient to give any clear insight as to the nature of their components, while opportunities are thus offered for acquiring much information about the paragenesis of minerals, the physical conditions under which rocks have been formed, and the changes both physical and chemical which they have subsequently undergone. It is therefore manifest that the preparation of thin slices of minerals and rocks has led the geologists of the present day into a vast and hitherto unexplored field of inquiry, in which new questions will propound themselves, and old ones, in time, find their solution.

The benefits to science which are likely to accrue from a steady prosecution of such studies are, from the foregoing remarks, sufficiently obvious, but many people may be prone to think that little practical advantage is likely to be derived from this branch of microscopic research. Such objectors, if asked whether the texture and quality of building stones, bricks, and mortars were matters of consequence in architectural work, would probably reply in the affirmative, and would then most likely add: 'Such matters can be decided by experience, by noticing buildings both old and new, and seeing how the different materials of which they are built have resisted the ravages of time and exposure, and if we want further information we can have the materials analysed by a chemist, and he will be able to tell us all we need to know.' Such a remark as this would embody a great deal of truth. The chemist could furnish an analysis of the stone itself, and we should thus learn how much lime, or magnesia, or silica, or carbonic acid, &c., &c., was contained in it; but if the stone happened to be a very fine-grained one, and although in freshly quarried samples apparently homogeneous, yet when exposed to the weather it suffered unequal decomposition or disintegration, it would be clear that atmospheric agencies detected weak spots better than the chemist, and better than the practised eyes of geologists, architects, stonemasons, and quarrymen. It would then be evident that more practically useful information could be derived from inspection of buildings than from the sources just enumerated; but at last the awkward question arises, 'How is it that stones which often have the same, or almost the same, chemical composition, and which also closely resemble one another in appearance, have different powers of resisting the effects which result from exposure?' It is true that the stones most affected may not have been judiciously laid; it is true that the atmosphere of large towns is more prejudicial to building stones than the purer atmosphere in country places; but, given similar conditions, why is it that apparently similar stones wear differently? Those who have examined the minute structure and mineral composition of rocks well know how little dependence is to be placed on outward similarity, and even in chemical analysis; for changes go on within rocks which often produce little or no definite chemical change so far as the aggregate is concerned, but beget numbers of little interchanges in the chemical composition and molecular arrangement of the component minerals. Although in the present state of our knowledge but little practical use has yet been made of the recorded observations which now constitute merely the small nucleus of what will no doubt eventually become a huge pile of information, still we may look forward to the time when a knowledge of the minute structure of rocks will be recognised as indispensable to the right understanding of the changes which building stones undergo, and when not merely the few but the many will be benefited by this branch of scientific inquiry. A general knowledge of petrology will always be found useful by those who may have to deal with architecture or with mining enterprises, and it is to be hoped that some day, as science progresses, a definite connection may be found to exist between metalliferous lodes and the mineral composition of the rocks in which such lodes occur. Questions of water supply hinge mainly upon the porous or impervious character of rocks, upon their mode of occurrence, and upon the structural planes or planes of dislocation by which they are traversed, so that matters of this kind can be best dealt with by the field geologist. In concluding this brief introduction, it seems needful to caution the student not to regard petrology from a narrow point of view; not to confine his attention solely to observations in the field; nor to devote himself exclusively to microscopical or chemical research. The disadvantage under which the specialist labours is, that he frequently takes infinite trouble to unravel a question in his own special way, when by adopting some other method he might arrive at his result in far less time, and often with greater certainty. At times a penknife will be found more useful than a blowpipe, and a blowpipe than a microscope; at other times a microscope will tell more than a complete chemical analysis.

CHAPTER II.

ROCKS DEFINED AND THEIR ORIGIN CONSIDERED.

IF we examine a fragment of rock, we find it to consist as a rule of crystals, the edges and angles of which may either be sharply defined or rounded, and which are cemented together either by crystalline or amorphous mineral matter, or we may find it composed of large or very minute angular fragments of mineral matter, or of rounded grains, or of a mixture of both angular and more or less rounded grains also bound together by mineral matter, which may either be amorphous or may possess a crystalline structure. Sometimes, however, the grains simply cohere without any perceptible cement, as in some of the new red and other sandstones. These kinds of rocks may be defined as mineral aggregates, but the term 'rock' in its geological signification does not merely imply a coherent mass but also loose incoherent mineral matter, such as blown sand, and in these cases, as in the preceding ones, the materials, whether they consist of fragments of one mineral only or of several different minerals, may still be regarded as mineral aggregates. There are, however, apparent exceptions, for some rocks appear to the naked eye to be perfectly homogeneous. Some quartzites may be so regarded, but we know that passages have been observed between quartzites and fine grained sandstones. A casual observer might also take such a rock as Lydian-stone or Hone-stone to be quite homogeneous, but examination of a thin slice under the microscope would show it to consist of numerous lenticular particles, which, from their form alone, imply the necessity for and the consequent existence of a cementing matter which differs from the particles themselves. Again obsidians, pitchstones, and other vitreous rocks, would be assumed by the general observer to be perfectly homogeneous, but here again the microscope demonstrates that they contain fine dusty matter, microliths, and crystals in great quantity; so that even those rocks which are apparently exceptions to the general definition are found in reality to conform to it, and thus we may with considerable truth define all rocks as mineral aggregates.

The state of aggregation of a rock depends upon the way in which it was formed and the changes which it may subsequently have undergone. We do not know what was the character of the rock or rocks which formed the crust when the surface of the heated mass which originally constituted the globe first solidified. The older geologists were fond of speculating upon this subject, and many of them believed that the primeval crust was granite, and they furthermore believed that all granite could claim this high antiquity. These were, however, mere speculations, and granites are now known to be of various ages, some of them having been formed in comparatively late geological times. Great misconception seems to exist among geologists even at the present day about the origin of rocks, and these misconceptions usually resolve themselves into bickering about terms and the way in which they should be employed. It therefore seems desirable, at the very outset, to lay before the student a few preliminary considerations which may help him to think and work in a systematic manner.

(1) Assuming that the earth was originally a molten mass revolving in space, and that after considerable radiation had taken place solidification of the surface ensued, it is clear that that primitive crust was the first rock formed.

(2) Assuming that radiation has constantly been going on from the earliest times up to the present day, it is evident that the whole mass of rock which now constitutes the crust of the earth represents the entire work of solidification, *no matter what may be the character (whether sedimentary, eruptive, &c.) of those rocks now*, unless, through changes in the interior of the globe, chemical action has been called into play, and has thereby generated additional heat, neutralising to some extent the work of solidification by fusing again the rocks which came in contact with these highly heated masses,

or, unless through fracture and disturbance of the already solidified crust (by the expansive force of gases generated by chemical changes, or by physical changes, such as the conversion of water into steam), portions of the already solidified crust were faulted down or depressed so as to come within the range of these heated portions of the earth, in which case, as in the preceding, some of the pre-existing work of solidification, from loss of heat, by radiation, would be undone, and the total mass of solid rock now constituting the crust would then represent the total amount of rock formed by solidification through radiation, minus the amount of rock re-fused.

(3) The rocks now called plutonic, and which have solidified at various depths beneath what was the surface of the earth at the time of their solidification, may be regarded as the result of the solidification of the earth's magma through loss of heat by radiation; the rocks now called volcanic, and which have been erupted through the crust and solidified on what was once, or on what is now, the surface of the earth (whether subaerial or submarine), may be regarded merely as outwardly trending phases of the plutonic rocks, and are therefore also the products of solidification through loss of heat by radiation; the rocks now called aqueous or sedimentary are the result of the degradation of land by subaerial or marine denudation, and because the first land must of necessity have been portions of the crust formed by the solidification of the earth's surface through radiation. Lastly, the rocks now called metamorphic are merely either eruptive (*i.e.* plutonic and volcanic rocks) or else sedimentary rocks, which have subsequently undergone changes either physical or chemical, and are therefore only the products of the solidification of the earth's magma through radiation, in an altered condition. Consequently no sedimentary rock has ever been formed except from materials which, in the first instance, were supplied in a solid form through the radiation of heat from the globe.

CHAPTER III.

DISTURBANCES OF THE EARTH'S CRUST.—STRUCTURAL PLANES.—SEDIMENTARY ROCKS.—STRATIGRAPHY.

THE study of forces existing in the interior of the earth, and the phenomena attendant upon their exertion and affecting the earth's crust, constitute the branches of physical geology known as vulcanicity and seismology, the former relating to volcanic phenomena and the latter to earthquakes. These forces are doubtless due to chemical and physical reactions and changes, resulting in the development of intense heat and the generation of gases. It is also assumed, however, that this high subterranean temperature is mainly owing to the original heat of the globe when first developed as an individual molten mass, only part of this heat having been dissipated by radiation into space. The consequent loss of heat having taken place from the exterior of the globe, and most affecting the superficial portion of it, has given rise to the formation of its crust, by solidification of the once molten matter. It seems reasonable to suppose that such radiation took place equally over the entire surface, and that the phenomena consequent upon the cooling of the mass, although doubtless affecting the centre of the globe, diminish in their intensity from the surface inwards. These physical changes probably extend inwards towards the centre with a certain amount of regularity, so that the globe might possibly be regarded as a spheroidal mass consisting of a series of zones varying in temperature and augmenting as the central portions are reached. Assuming this to be the case, there would be a zone situated at some depth beneath the surface, whose temperature would be so high that any known rock matter could no longer exist in a solid state. The depth at which this zone of fusion is supposed to occur has been variously estimated by different writers. From twenty-five to thirty miles is about the smallest estimate which has been given; while some, as Hopkins, have inferred that solid matter extends to much greater depths and may even exist at the centre, the loss of heat and consequent solidification having taken place in an irregular manner, and having thus converted the deeper portions of the globe into a somewhat honey-combed mass, the cavities still retaining matter in a molten condition, and constituting the reservoirs from which the eruptive rocks are derived. So many theories upon this subject have from time to time been started, and they embody such diversity of opinion, that a description of them would be out of place in so small a book; and as they cannot be regarded as more than speculations, often based upon a little tangible truth and more or less tangible and intangible error, students, although doing well to make themselves acquainted to some extent with these theories, would do better in giving their attention to matters which are more readily demonstrable.

That the forces just spoken of as existing in the interior of the earth exercise considerable influence upon its crust we have ample evidence. It is, indeed, solely from such evidence that we infer the existence of the forces.

The evidence which we have of the internal heat of the earth may be summed up under the following heads:—

(1) In descending the shafts of mines a gradual rise of the thermometer takes place after the descent of the first sixty feet. Down to this point it remains stationary; below this point there is a rise of one degree Fahrenheit for every sixty feet descended. How far this regular increase of temperature continues to take place has not yet been determined.

(2) Flows of molten lava and of hot mud, the ejection of lapilli and ashes from volcanic vents through the generation of steam or the evolution of gases, and the occurrence

of geysers and thermal springs, are also evidences of the internal heat of the earth.

(3) Earthquakes, the elevation and depression of large tracts of land, giving rise to changes in coast lines, as in Greenland, Sweden, and South America, at the present day; the fractures produced in the crust by subterranean forces, and the relative displacement of rock masses along such lines (faulting); the bending and contortion which stratified rocks undergo, and the chemical and physical changes (metamorphism) which they sometimes experience, are all evidences of the internal heat of the earth.

That the fissuring and displacement of portions of the earth's crust is frequently due to the exercise of the subterranean forces just mentioned there is not the least doubt, since they have occurred frequently, not merely within historical times, but men now living have been eye-witnesses of many remarkable changes which have been brought about in the structure of districts by seismic action, as in Calabria and in the country at the mouth of the Indus.[1] Moreover, many of the disturbances of the earth's crust which have happened in even very remote geological periods are precisely such as we should attribute to the same forces which have produced, and are still producing, the modern disturbances. We may conveniently classify them all under certain heads, so far as they relate to structural planes; either on a large scale, as affecting the general geology of a country, or on a smaller scale, as influencing not merely the gross but the minute and even microscopic structure of rock masses. It may here be remarked that the scenery and general configuration of a district is often due rather to the facilities offered for the weathering of rocks along small and closely disposed planes of fission than to the presence of long lines of fracture and faulting. The latter tend to produce

[1] For descriptions of these and other kindred phenomena the student should consult Lyell's *Principles of Geology*, in which accounts are given of the most important earthquakes and volcanic eruptions which have occurred within historical times.

lithological diversity of surface rather than diversity of contour or relief; atmospheric agencies apparently producing little or no effect upon one colossal divisional plane, since a fault seldom developes a feature in any landscape, other than perhaps a difference in its vegetation. Atmospheric degradation, however, along innumerable divisional planes of very trivial dimensions gives rise to outlines which often enable a practised observer to discriminate between different formations merely from the aspect which they have derived from weathering.

The following table gives a rough and rudimentary classification of these structural planes:—

CLASSIFICATION OF STRUCTURAL PLANES.

1. Irregular fissures and faults.
   - *Earthquakes and their attendant phenomena*, producing fracture of the crust by pressure from within directed outwards.

2. Jointing due to
   - *Shrinkage on consolidation of sediment by drying and consequent contraction*, producing fracture of the crust usually along more or less parallel lines.
   - *Shrinkage on consolidation of eruptive matter by cooling and consequent contraction*, producing fracture in directions more or less parallel.

3. Laminar fission and cleavage.
   - *Pressure, exerted by contiguous rock masses*, producing (often by re-arrangement of particles) planes of weak cohesion, along which fission readily takes place in parallel directions.
     - A. Coincident with bedding planes (laminar fission or flaggy cleavage).
     - B. Deviating from the direction of the bedding planes (slaty cleavage).

It is possible that many of the lines along which faulting has taken place may in the first instance have been simply fissures due to shrinkage (Class 2) but in other cases faulting has occurred along planes produced by seismic action (Class 1).

The parallelism which so often characterises a system of faults cannot, however, be adduced as proof that those faults have occurred along joint planes (Class 2), since parallel fissures might be produced by the upheaval of rocks along a certain line or rather along a definitely trending area (Class 1) on either side of which relative displacement of strata or of eruptive rock masses may have subsequently occurred.

It is also quite possible that the displacement itself originated synchronously with the line or plane along which it runs, in which case the plane would again belong to Class 1. All that we can therefore safely say about the origin of faults is, that they are relative displacements of the earth's crust, caused by subterranean forces upheaving masses of rock along lines of least resistance, which may either be produced at the time of upheaval or may have pre-existed simply as fissures or cracks, and that in some cases depression of rock masses has caused faults, the subsidence of the downthrow having occurred by the mere gravitation of the mass between two outwardly diverging planes of fracture, as in the case of 'trough faults.'[1]

Faults may also arise from an unequal horizontal shifting of undulating beds along a fissure, or from partial flexure or bagging down of strata upon one side only of a fracture, the beds on the other side remaining horizontal.[2]

The fissures and cracks, therefore, along which faulting has taken place, may be due either to volcanic or seismic action (for earthquakes and volcanic phenomena are so intimately related, and are apparently so indisputably due to similar or identical causes, that they may safely be classed together), or to the shrinkage of sedimentary rocks by loss of

---

[1] Vide *Student's Manual of Geology*, by J. Beete Jukes (1862), p. 260.
[2] *Ibid.* pp. 254–5.

moisture and that of eruptive rocks by loss of heat during solidification.

Another class of small structural cracks which occur in some eruptive rocks and also exceptionally in argillaceous beds which have undergone considerable desiccation, is, in the opinion of many observers, due, in the case of the eruptive rocks, such as basalt, to contraction on cooling. These planes intersect one another in such a manner as to divide the mass of rock into a series of closely packed prisms varying at times in the number of their sides and in the measurement of their angles. This structure is especially characteristic of the basalts in some districts (Staffa, Giant's Causeway, Unkel on the Rhine, parts of Auvergne, and many other localities.) The prisms are generally cut transversely by numerous divisions, which are sometimes flat, sometimes either convex or concave, while occasionally, as in the celebrated Kasekeller, they consist of superposed spheroidal lumps or balls, which have a concentric shaly structure. An analogous structure on a very small and often purely microscopic scale is to be met with in vitreous rocks such as perlite.[1] According to the theory held by Sir Henry De la Beche and others, mountain chains owe their origin in many cases to immense fractures and dislocations of the earth's crust caused by unequal contraction of the crust in zones, the inner zones contracting and leaving the outer and already solidified zone unsupported, so that in places it cracked, large masses subsided on to the lower zone, and thus caused immense ridges and depressions. Such mountain-forming fissures, colossal though they may be, are, however, hypothetical rather than demonstrable.

Enough has now been said to show that structural planes and divisions occur in rocks ranging from those of gigantic size to others of quite microscopic dimensions. Some of them occur in rocks of eruptive origin, some traverse sedi-

---

[1] Bonney, *Q. J. G. S.*, vol. xxxii. p. 140; Allport, *Q. J. G. S.*, vol. xxxiii. p. 449; Rutley, *Trans. R. Mic. Soc.*, vol. xv. p. 176.

mentary deposits, but in all cases they facilitate the weathering and disintegration of the rocks in which they occur, and consequently exercise a more or less marked effect upon the scenery of a district.

The rocks composing the crust of the earth may be considered mainly to belong to two great divisions, viz., (A) the aqueous, sedimentary, fossiliferous, or stratified rocks, which have been deposited as sediment in beds, or strata beneath water, each bed or stratum having successively formed the floor of a sea, or of a lake; and (B) igneous or eruptive rocks, which have formed intrusive bosses, or dykes, or have been poured out from volcanic vents, as lava flows.

The former usually contain organic remains which may be identified with a marine or a lacustrine fauna, and consequently afford a tolerably safe clue to the circumstances under which the beds were deposited. It may be safely assumed that all such beds were originally spread out in an approximately horizontal position, and that any strong deviation from the horizontal position which may be shown by planes of bedding is due to subsequent disturbance of those beds.

Sediments may sometimes, however, be somewhat irregularly deposited; for example, a number of thin beds may thin out completely, overlie one another, and the whole of them may overlie a perfectly horizontal bed upon which their thinned-out ends appear to rise unconformably, and this kind of arrangement may be repeated again and again through a considerable thickness of deposits. This irregular kind of stratification is called 'false bedding.' Fig. 1 represents a good example, occurring in the lower greensand, at Frith Hill, near Godalming.

In such a case the inclination of the beds is not due to any disturbances during or subsequent to deposition, but simply to the overlap of successive deposits as they are thrown down in shallow water. The sediments which constitute stratified rocks result from the wear and tear which takes place from the action of rain on land surfaces, and, in the beds of rivers from attrition, each eroded fragment serving as a tool with which other fragments are ground away from the rock. When a turbid river empties itself into the sea or into a lake, the materials held in suspension become deposited according to their relative specific gravities, the heavier fragments sinking first, while the lighter particles are carried to a greater distance from the

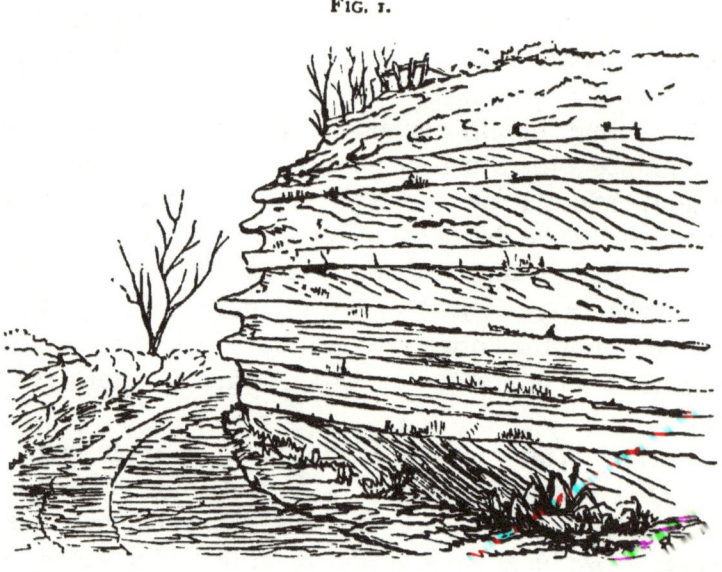

FIG. 1.

river's mouth. The relative sizes and shapes of the fragments also exercise some influence on the sorting process which takes place. Fragments which have undergone but little attrition are usually more or less angular in form, while those which have been carried long distances, and which have been rubbed together for a length of time, become sub-angular or perfectly rounded.[1] The rounded form of the pebbles on sea-beaches is due to the incessant grinding

[1] Except in instances where the fragments have been transported by ice.

which they undergo against one another during the advance and retreat of every wave that washes the shore.

Deposits mainly composed of angular fragments are termed breccias, while those consisting of rounded pebbles are called conglomerates. In indurated rocks of this kind the coarse fragments, or pebbles, are generally cemented together by a finer material, often consisting of carbonate of lime or silica. There are frequently other substances, however, which act as a cement.

Aqueous or sedimentary rocks may be conveniently classed as follows:—

(1) *Clays*, which when indurated become *mudstones*, and, when cleaved, *slates*. When they merely exhibit a fissile character in the direction of the lamination, or bedding, they are called *shales*. Clays, slates, and shales are mainly composed of hydrous silicate of alumina. There are arenaceous and calcareous clays, slates, and shales; calcareous clays are termed marls.

(2) *Sands.*—These when indurated constitute *sandstones*, and when more or less coarse-grained, and composed of angular or sub-angular grains of sand (frequently with an admixture of fragments of other minerals), they are then termed grits. The term 'grit' is, however, very loosely used, and it would be difficult to give it a sharp definition owing to the great variation in the physical and mineralogical characters of the rocks to which this name has been applied. Sands and sandstones are usually composed of fine grains of quartz cemented either by carbonate of lime, carbonate of iron, oxides of iron, or silica. There are calcareous and argillaceous sandstones.

(3) *Limestones.*—These may vary from soft and earthy, to hard, compact, and even finely crystalline rocks. Some limestones may be merely eroded granules of pre-existing limestone carried mechanically in suspension in water, and ultimately deposited as a sediment. Some may have resulted from the precipitation of carbonate of lime from water holding the bicarbonate of lime in solution. In this

case the deposit may be considered to have a chemical origin. The cause of precipitation would be the elimination of one atom of carbonic anhydride from each molecule of bicarbonate of lime. Travertine, calcareous tufa, and pisolite are rocks formed in this manner. The last consists of rounded grains like shot or peas, whence the name pisolite or peastone; these little pellets consist of a series of concentric coats of carbonate of lime which sometimes have a small grain of sand as a nucleus. Limestones are also at times composed in great part of the shells of minute animals called 'foraminifera.' These organisms, whose remains constitute the very earliest record of life of which we have any knowledge, have peopled the waters of various geological epochs with their descendants, and at the present day the foraminifera have numerous living representatives. The animals themselves are little more than small shapeless masses of animated jelly, but they have the power of separating carbonate of lime from solution in water, and of building up the material into shells of very variable and extremely beautiful forms. Some are perforated by immense numbers of minute holes through which the gelatinous occupants can protrude their filamentous processes. To these holes, or foramina, the order owes its name. Corals also have the power of secreting large quantities of carbonate of lime, and some limestone rocks are in great part due to the secretions of these polyps. The shells of the mollusca, which have originated from a similar secretive faculty, also at times contribute largely towards the formation of some limestones. This secretive process can merely be regarded as a chemical process performed through the intervention of the animal; and when we speak of such rocks as having an *organic origin*, we must be careful not to imply that the animal had actually formed the calcareous matter instead of having merely secreted it. Pure limestones consist simply of carbonate of lime. A compound of the carbonates of lime and magnesia constitutes magnesian limestone, or dolomite. Those limestones which contain a certain amount of clayey matter are termed argillaceous limestones, and those containing sandy impurities are styled arenaceous limestones.

The changes which sedimentary rocks undergo may be regarded as physical, as chemical, or as the result of physical and chemical agencies acting either simultaneously or at different periods.

CLASSIFICATION OF THE SEDIMENTARY ROCKS.

ARGILLACEOUS GROUP.
- *Clay.* Composed of hydrous silicate of alumina, usually with mechanical admixture of sand, iron oxides, and other substances.
- *Marl.* Clay containing calcareous matter.
- *Shale.* Indurated clay, fissile along planes of bedding.
- *Slate.* Indurated clay, fissile in parallel planes other than those of bedding.

ARENACEOUS GROUP.
- *Sand.* Chemical composition silica. Mineral components quartz or flint.
- *Sandrock.* Coherent sand.
- *Sandstone.* A more or less strongly coherent and often highly indurated sand.
- *Grit.* A coarse-grained and somewhat coherent, or at times a fine-grained and very hard and compact sandstone, frequently containing fragments and granules of other minerals beside quartz, flint, or chert.
- *Calcareous Sandstone.* Sandstone cemented by carbonate of lime.
- *Ferruginous Sandstone.* Sandstone cemented by an oxide of iron or by carbonate of iron.
- *Conglomerate.* Rounded pebbles of flint, chert, jasper, quartz, &c., cemented either by siliceous calcareous, or ferruginous matter.
- *Siliceous Breccia.* A rock similar to the above in composition, but differing from it in containing angular fragments instead of rounded pebbles.
- *Siliceous Sinter.* Silica deposited in a more or less loose or spongy form from waters holding silica in solution.

Many of the slates, sandstones, and grits afford good building stones.

CALCAREOUS GROUP.

Limestone. Chemical composition, carbonate of lime. Limestones vary greatly in their physical characters; some being earthy, soft, and friable, as chalk; others hard and crystalline.

The principal limestones used for building purposes are the Devonian, the carboniferous, and the magnesian limestones; many of the oolitic limestones, especially the Bath stone, Portland stone, and Purbeck limestone.

Limestones which are capable of receiving a polish are called marbles. They vary so greatly that it is not possible to describe even the leading kinds in a small space.

Bands of chert occur in the carboniferous and in some other limestones, as the Portland, and bands and nodules of flint are met with in the upper chalk.

Magnesian Limestone. Chemical composition, carbonates of lime and magnesia. This rock is also called Dolomite, after Dolomieu. The proportions of carbonate of lime to carbonate of magnesia vary greatly in different localities.

Argillaceous Limestone. Limestone containing some clayey matter or hydrous silicate of alumina. When this reaches a certain proportion the rock is termed an hydraulic limestone; such limestones are used for the manufacture of cements which set under water (hydraulic cements). The lias limestone is a good example of an argillaceous limestone.

Arenaceous or Siliceous Limestones represent a transitional condition between limestone and chert. Some of them, such as the Kentish rag, afford good building stones.

It has been already stated that the sedimentary rocks occur in beds or strata (hence they are also called stratified rocks). This arrangement has, in the first instance, been an approximately horizontal one, and, in most cases, where there is any marked deviation from horizontality, the deposits have been disturbed by the action of subterranean forces. When any such disturbance has taken place, so as to communicate an inclination to the beds, this inclination is termed 'dip.' If we assume a long strip of paper to represent a horizontally deposited bed or stratum, and then fold it lengthwise as in fig. 2, we have two dips, one in the direction of a and the other in an opposite direction, b.

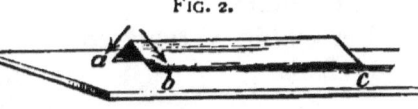

Fig. 2.

The direction bc or cb is termed the 'strike' of the beds. The strike is always an assumed horizontal direction, so that if we tilt our strip of paper on one end the strike will still be a horizontal line as de (fig. 3). The dip is always reckoned at right angles to the strike. It is somewhat difficult to

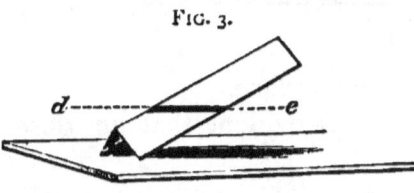

Fig. 3.

render this apparent in a diagram; but if we represent one side of our strip of paper to be dipping vertically, *i.e.* at 90°, as in fig. 4, it will render the mutual directions of dip and strike as seen in plan: ab representing the strike, and cd the dip.

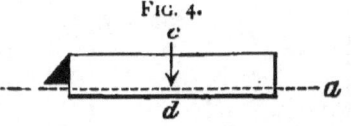

Fig. 4.

In nature it is not usual to find beds bent in the acute manner indicated in the preceding figures. They generally describe curves which represent the arcs of circles sometimes many miles, sometimes only a few feet in extent. In the latter case this small crumpling is spoken of as 'contortion.' When flexure of strata occurs in an upward direction the result is spoken of as an anticlinal flexure, curve, or ridge; while, on the other hand, when the curve is directed downwards in a basin-shaped manner, it is termed 'synclinal.'

The dip of strata exercises a marked influence on the scenery of a country. If no disturbance of stratified deposits had ever taken place in a district no knowledge of its geology could be obtained, except along valleys which had been scooped out by the action of rain, rivers, and general atmospheric agency, or in railway cuttings, quarries, and other excavations, and in borings such as wells and the shafts of mines.

Let A B (fig. 5) represent the surface of a country in which the strata have never been disturbed, and therefore lie horizontally just as they were deposited. Let us also suppose that the surface has not been carved out into hills and valleys, but is a level, unbroken surface.

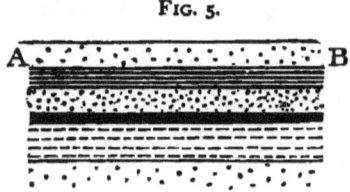

FIG. 5.

It is evident that an observer walking across such a district would meet with no diversity in the character of the soil or of the rocks over which he passed, unless indeed the same deposit exhibited slight lithological change in its own horizon—such as a passage from clay into sandy clay, and the geological map of the district represented by the section A B would, if coloured, be merely painted over with one uniform tint, or stippled thus (fig. 6).

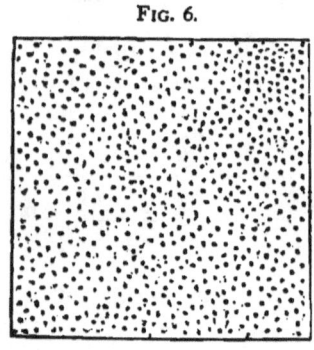

FIG. 6.

If such a country were scooped out by the action of rain, rivers, &c., the section A B (fig. 5) would undergo considerable alteration, as shown in the accompanying fig. 7;

while its geological map would now indicate the exposure not merely of the uppermost stratum, but also the outcrop of several underlying beds somewhat in the manner shown in fig. 8, with probably a river R R running along the bottom of the valley.

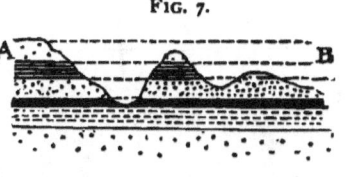

Fig. 7.

Maps of the oolitic and liassic districts of England will be seen to resemble this in their geological boundary lines, and these lines usually follow the general contours of the country.

Fig. 8.

Let A B (fig. 9) represent the surface of a country in which the stratified rocks have been disturbed and tilted upwards; in other words, a country in which the strata have a definite dip.

It is highly improbable that such a country would have a level surface, as the unequal hardness of the different rocks exposed to the action of the atmosphere would tend to beget considerable irregularity, the harder ones being worn away less easily than the soft ones.

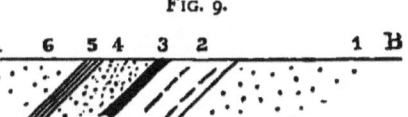

Fig. 9.

Still, supposing the surface of the country to have been planed off to a perfect level, the geological map of the district would nevertheless present great diversity in its colouring or shading. A man walking across the country from east to west would pass over several different formations. In this case s s, s s (fig. 10) would represent the strike of the beds, and the arrows would show the direction of their dip, at right angles to the strike.

If the beds were repeated, dipping in an opposite direction, *i.e.* if they had an anticlinal arrangement, some estimate could be formed of the amount of rock which had been denuded by restoring the curves as indicated by the dotted lines in fig. 11, although this would probably represent only a portion of the total amount of matter which had been removed.

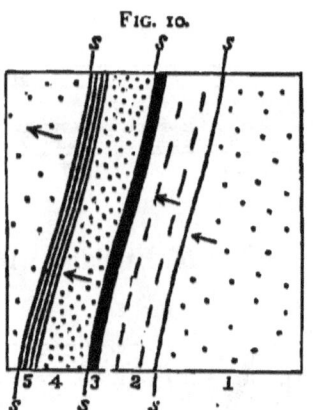

Fig. 10.

The thickness of beds should in all cases be measured at right angles to the planes of bedding, whether they be undis-

Fig. 11.

turbed and horizontal or disturbed and inclined. Thus, if A A (fig. 12) represent the surface of the ground, and B an inclined bed, then the thickness of B should be measured along the dotted line X X, and not along the surface A A.

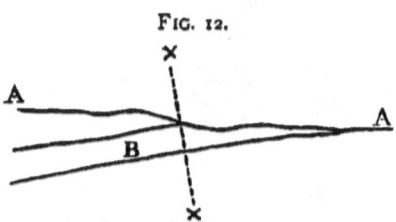

Fig. 12.

Enough has now been said to denote the way in which the disturbance of sedimentary rocks influences the surface of a country, and, when aided by denudation, promotes our knowledge of its geology, by bringing to view subjacent deposits which would otherwise have been accessible only by excavation and boring, thus affording us the means of selecting from various sources materials of industrial importance—such as building stones; bringing within workable distance various mineral deposits, and diversifying the surface of the land in a manner which affects agriculture and water-supply, civil engineering, and last, but not least, the sanitary condition of its inhabitants.

The manner in which rocks are worn away is spoken of as denudation. Denudation may be regarded under two heads:

(1) Marine denudation.
(2) Atmospheric or subaerial denudation.

The tendency of all denudation is to wear away existing land to lower levels until it reaches the level of the sea. The wearing process then stops, because the sea can only act destructively in planes situated between high and low water-mark. The breakers do all the work of marine denudation; and when they can no longer act upon rocks because they have planed them down so far as they can plane them, that is, to their own level, the process of denudation of course ceases; and when such a stage of degradation is reached, subaerial denudation also becomes inert. Marine denudation may therefore be defined as the degrading influence exercised by the sea on a level with the breakers. This degradation is effected not merely by the force of the breakers dashing and pounding against their barriers, but the detached fragments are hurled again and again against the rocks, which thus continually supply ammunition for their own destruction. The tendency of all this is to cut back the coast into cliffs; and the process, so far as the sea is then concerned, becomes an undermining one. This is well shown in Shakespeare's Cliff at Dover, where the sea has undercut the chalk at the base of the cliff (fig. 13); but it is usual to see few signs of excavation at the foot of a cliff, because subaerial denudation is also continually going on; and when its effects upon certain rocks are more marked than those of marine denudation, the face of the cliff will be seen to slope backwards more or less.

If marine denudation acted more powerfully than atmospheric denudation, then the cliffs would become undermined, and the overhanging masses would eventually give way, falling on to the shore and forming a talus, or heap of broken fragments banked up against the lower portion of the cliffs. A talus may also be formed by the fall of fragments dislodged by frost and other atmospheric denuding agents, so that under any circumstances it is common to find the foot of a cliff so protected, except where the shore is very steep or where the scour of the tide is considerable.

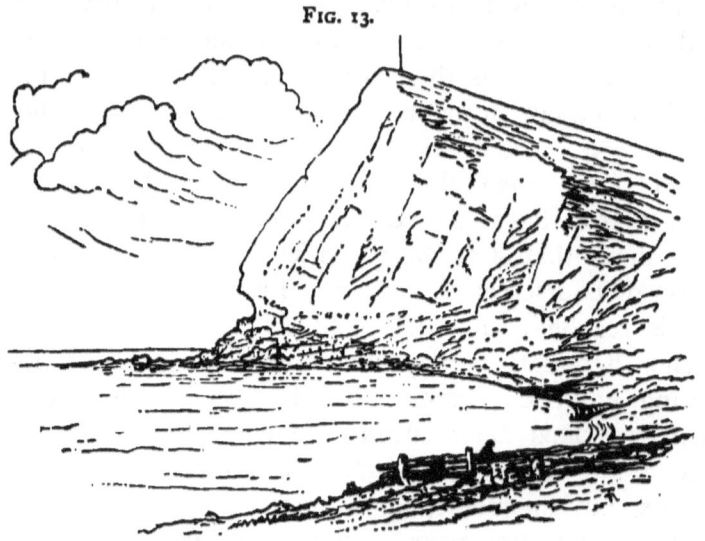

Fig. 13.

These natural barricades prevent the sea from attacking the base of the cliff for a time, but after a while they are cleared away and the undermining process recommences, to be again temporarily retarded by successive falls of rock from above. When the stratification of the rocks which form a coast dips inland, the sea acts very destructively, large fragments being dislodged with comparative ease; but when the dip is seawards, the breakers run up inclined planes of bedding instead of dashing against an abruptly raised barrier, and so the work of degradation takes place much more slowly. Sometimes steep slopes, forming features which to a certain extent resemble cliffs, are seen far inland. These are called escarpments. They owe their origin to subaerial

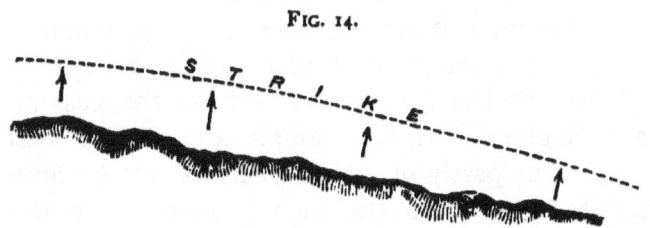

FIG. 14.

and not to marine denudation, and they differ from true sea-formed cliffs in that the escarpment will be found to trend in a direction parallel to the strike of the beds (fig. 14),

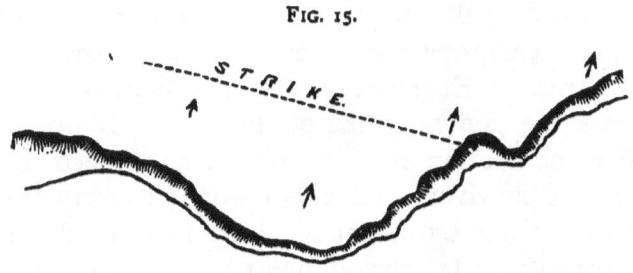

FIG. 15.

while a coast line backed by cliffs runs in an irregular manner, which bears no relation whatever to the strike (fig. 15). It is therefore evident that the configuration of a country, so far as its coast-line is concerned, does not depend upon the strike of the rocks along that coast-line except so far as the strike determines the position along that coast-line, of relatively hard and soft rocks,

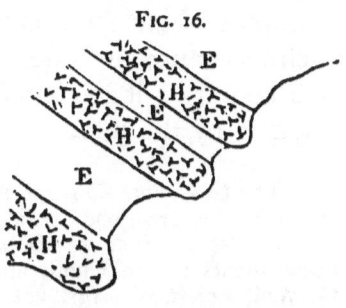

FIG. 16.

or of rocks whose chemical or physical characters render

them relatively difficult or easy of disintegration, as in fig. 16, where H represents rocks comparatively difficult, and E those which are comparatively easy, to wear away.[1]

The difference in the weathering of rocks, dependent upon their relative hardness, does not merely influence the coast-line of a country, but it affects its inland configuration to a greater or less extent, as shown on the east and west coasts of England. If, for example, a country consist partly of granite and partly of slate, it will usually be found that the granite constitutes the high ground, while the slates occupy the lower portions of the district. It does not, however, necessarily follow that all the valleys of a country should be scooped out in the softer rocks, while the harder ones only form the hills. If this were the case, the drainage of a country would be determined by the general strike of the rocks, and all the valleys would trend in the direction of the strike. This, however, is not always the case. In the Lake District of England, for example, most of the valleys run across the general strike of the Upper Silurian rocks, instead of coinciding with it. It seems, indeed, that the directions of the rivers and valleys of a country are often determined rather by its initial slope than by the relative resistances offered to erosion by its rocks. As a rule, however, it is probable that neither of these considerations can be utterly ignored, and that the truth involves them both : first, the initial slope of the district ; and secondly, the relative resistance which the rocks offer to atmospheric denudation. Such resistance does not merely influence the large features of a country ; it renders itself evident in the actual shapes assumed by the hills.

[1] The *Geological Observer* and *How to Observe Geology*, by Sir Henry De la Beche, are good works for the student to consult upon these points; also a Paper on 'Subaerial Denudation and on the Cliffs and Escarpments of the Chalk and the Lower London Tertiary Beds,' by W. Whitaker, *Geol. Mag.* vol. iv. pp. 447–483. London. 1867. The phenomena of denudation are also well described in Prof. Ramsay's *Physical Geology and Geography of Great Britain*.

The forms of hills and mountains are mainly due—

A. If composed of sedimentary rocks—

(1) To the lie of the beds, whether horizontal or inclined.
(2) To the presence or absence, the paucity or multitude, of the structural planes which traverse those beds.
(3) To the physical characters of the beds.
(4) To their chemical composition.

B. If composed of eruptive rocks—

(1) To the presence or absence, paucity or profusion, of structural planes.
(2) To the physical character of the rocks.
(3) To their mineral constitution.
(4) To their chemical composition.

The average rainfall of a district of course has more or less influence on the weathering and disintegration of rocks, while the water which filters into them, especially along joint planes and fissures, expands, on its conversion into ice, in frosty weather, and greatly facilitates their degradation, by forcing apart fragments and blocks of rock which become detached as soon as the ice thaws. The process of disintegration is carried on to a great extent by the rain, which, in its passage through the atmosphere, absorbs a considerable quantity of carbonic acid. This acts upon compounds of lime, potash, soda, &c. In limestone districts a great amount of matter is annually removed in this way by the conversion of the carbonate into the soluble bicarbonate of lime. The water of rivers, and the water from swamps and peat-mosses also, contains more or less carbonic acid, especially in the latter cases, where vegetable matter by its decomposition gives off considerable quantities of this gas. Limestones are by no means the only rocks acted upon by waters containing carbonic acid.

The felspars, which constitute so great a portion of many eruptive rocks, suffer decomposition through this cause. If the felspar be a lime-felspar the lime is the substance first removed; the potash and soda, which most felspars contain, are next carried off as carbonates, although they are sometimes removed in the form of soluble silicates. This gradual removal of the alkalies, &c., ultimately results in the formation of kaolin or china-clay, a hydrous silicate of alumina, $\overset{..}{Al}\overset{..}{Si}_2 + 2\overset{.}{H}_2$. Should the waters contain magnesian salts in solution, the lime or soda of the felspars may be replaced by the isomorphous magnesia, thus giving rise to steatitic matter so long as the alumina of the felspar is not involved in the change. Through similar causes the various species of mica become at times partially decomposed and converted into steatite, serpentine, compounds allied to chlorite, and possibly some other hydrated minerals. The alterations and replacements of minerals, by the infiltration of water charged with acids and soluble salts through the rocks in which they occur, are so numerous, and the pseudomorphs to which they give rise are so interesting, that special mention of them will be made in those parts of this work where the rocks in which they are found are described. The interchanges which take place in the formation of these pseudomorphous minerals are in great part due to the isomorphism of the replaced and the replacing substances, although at times the original mineral may be entirely removed and its place subsequently occupied by matter bearing no such relation to the components of the mineral replaced.

It has already been stated that the sedimentary rocks occur in layers, technically termed strata or beds. A number of such beds, deposited at the same time, under approximately the same conditions, in one or in many distinct areas (areas which during deposition may once have been united, or in which deposition may have gone on independently), are termed 'formations.' The formations are usually more or less fossiliferous, and separate formations are distinguished by characteristic fossils, which bear testimony to a certain sameness in the animal and vegetable life which existed from the deposition of the lowest to that of the highest beds in the formation. The geological age of any bed may therefore be determined either by its stratigraphical horizon or by the fossils which it contains. The lithological character of a bed sometimes varies, and, therefore, less dependence is to be placed upon it than upon palæontological evidence. Furthermore, the slight variation in the lithological characters of sedimentary rocks often renders it very difficult to assign them to any particular horizon in the absence of fossils; sandstones, slates, shales, and limestones of different geological ages often bearing a close resemblance to one another. Again, rocks differing widely in lithological character may have been deposited at the same time, as in the case of the Devonian and old red sandstone rocks, the former having been thrown down in the sea, and the latter in lakes, as proved by the fossils which they respectively contain. Yet both formations occupy a position intermediate between the Upper Silurian rocks and the lowest members of the carboniferous series. The grouping of sedimentary rocks into formations is, of course, more or less arbitrary. *Some* genera, and frequently species, which occur in a lower formation are often represented in the succeeding deposits of a newer formation, and probably, if the truth were known, it would be found that all the formations, which we now recognise, pass from one into another. Because an unconformity, *i.e.* a break both stratigraphical and organic, occurs in one limited district, it does not necessarily follow that this break extended over the entire globe. Allowances must be made for relative distributions of land and water, which we have often no means of realising, and no doubt the universal application of limited knowledge often does more harm than good in this branch of geological inquiry.

CHAPTER IV.

ERUPTIVE AND METAMORPHIC ROCKS.

THE eruptive or igneous rocks differ entirely from those of sedimentary origin in their mode of occurrence (except in the case of volcanic ejectamenta, presently to be explained, in the interbedding of lava-flows, and in the intrusion of sheets of eruptive rock between planes of bedding, as in the case of the Whin Sill of Northumberland[1]). They bear, except in such instances as those just cited, no definite relation to the sedimentary rocks, but form irregular masses often of very great extent, from which vein-like prolongations of tabular and wall-like masses (dykes) are often sent off into the surrounding rocks. They also emanate from volcanic vents in the form of molten viscous lava, forming flows or coulées, which are forced over the edge of the crater, frequently breaking it down on one side, creeping down the sides of the cone, and often spreading for many miles over the surrounding country.[2]

Sometimes they are erupted in the form of large and small fragments of rock (lapilli), of peculiar spheroidal molten masses (volcanic bombs), and of finely comminuted and dusty mineral matter (ashes). These fragmentary ejectamenta are often thrown high into the air. Part of them fall back again into the crater to be again and again thrown up, so that by constant attrition they become more or less rounded. Part fall on and outside the rim of the crater, thus helping to build it up higher. Part may be carried by

[1] *Vide* paper 'On the Intrusive Character of the Whin Sill of Northumberland,' by W. Topley and G. A. Lebour, *Q. J. G. S.*, vol. xxxiii. p. 406.

[2] Intrusive sheets may be distinguished from true lava-flows, which have been subsequently overlaid conformably by sedimentary strata, by the fact that the rocks both above and below the intrusive sheets are altered at the contacts, while in the case of lava-flows the rocks over which they ran have been altered, but the deposits above them show no trace of metamorphism.

the wind and showered down over the adjacent country, and if in the state of very fine dust may be transported immense distances by the wind, a passage of between 700 and 800 miles having been recorded. There are also some craters, usually of comparatively small dimensions, which pour out liquid mud, frequently accompanied by an outpouring of water also. The water is in some cases boiling, in others cold, and bitumen has also been seen to exude from some of them. The hot water ejected by the Geysers of Iceland, and that of the thermal springs of Roto-Mahana, near Lake Taupo, in New Zealand, carry a large amount of silica in solution, which on the evaporation of the water leaves a deposit or incrustation of white siliceous sinter. Besides the lavas, ashes, &c., which emanate from volcanoes, steam and gases, such as carbonic anhydride, sulphurous acid, hydrochloric acid gas, sulphur vapour, &c., are also emitted.

It is still customary to divide the eruptive rocks into two classes—the plutonic and the volcanic, the former class including those rocks which have solidified at considerable depths beneath the earth's surface, and which are now only exposed because the rocks which once overlaid them have been removed by denudation. The volcanic rocks, on the other hand, although likewise originating at considerable depths, have been forced up until they not merely reached but in many cases have overrun the surface. From the mode of occurrence of the rocks belonging to these two classes it is by no means easy to affirm with positive certainty that they are merely different phases of the same eruptive rock-forming matter, emanating from the same source. Still, on comparing the rocks of the one class with those of the other, a tolerably continuous chain of evidence can be adduced to show that they graduate into one another, and that this, like all other classifications, which are necessarily arbitrary, is more hypothetical than real. The quartz-porphyries or elvans resemble the granites more or less in mineral composition, and are known to emanate from granitic masses; but the mica, which is plentiful in granites, is only poorly represented or is totally absent in the quartz-porphyries. The porphyritic felstones resemble the quartz-porphyries, except that they contain no definite and well-marked crystals and blebs of quartz. The felstones which are not porphyritic are often identical with the magma of quartz-porphyry. The trachytes vary in their affinities, some (such as the quartz-trachytes) inclining more towards the granites and felstones in mineral composition, while others (as sanidine-oligoclase-trachytes) occupy an intermediate position, passing into or becoming allied to the basalts, dolerites, &c., in the oligoclase trachytes, the andesites, and the trachy-dolerites.[1] If, then, such close resemblances in mineral constitution can be discerned in the rocks which come near the boundary line drawn between the two classes, and since rocks containing under 60 per cent. of silica (basic, of Bunsen), and over 60 per cent. of silica (acidic, of Bunsen) are found in both classes, it seems reasonable to suppose that close resemblances in mineral constitution are almost equivalent to observed passages. Again, both acidic and basic rocks are known in some instances to have emanated at different periods from the same volcanic vent. Durocher, in the 'Annales des Mines,' vol. xi., 1857, enunciated the theory that all eruptive rocks have been derived from one or other of two magmas which occur in distinct zones beneath the solid crust of the earth, the one poor in basic materials, but containing over 60 per cent. of silica, the other rich in basic matter, but holding less than 60 per cent. of silica. The former magma having a less specific gravity than the latter, is assumed to float, as it were, upon it, the

---

[1] Mr. J. Clifton Ward states that some of the rocks occurring in the English Lake District are intermediate in mineral composition between felstones and dolerites, and in describing them he designates them 'felsi-dolerites.'—*Q. J. G. S.*, vol. xxxi. p. 417. In examining some of the eruptive rocks from the Silurian districts of North Wales the author has met with similar examples, and can fully endorse Mr. Ward's conclusions.

Cleavage. 35

difference in specific gravity of the rocks derived from these magmas being from one and a-half to twice as great as between oil and water. Taking all these things into consideration, we may be justified in assuming that the difference between the plutonic and volcanic rocks of Bunsen's acidic class lies wholly in the fact that they have solidified under different conditions, but that their differences do not sufficiently warrant their separation by a hard and conventional boundary-line, nor debar us from the inference that they may often have arisen from the same deep-seated sources. The same may be said of Bunsen's basic class. If it could be demonstrated that they *have* done so, nothing would remain but to admit that the plutonic rocks are the roots and stem, the volcanic rocks the branches and twigs, of a great petrological system.

The sedimentary rocks, as already mentioned, occur in beds, or strata which were originally deposited in an approximately horizontal manner. Furthermore, the beds generally exhibit lamination (or still finer bedding). The mineral particles of which these rocks are composed are, when inequiaxial, arranged with their longer axes parallel with the lamination or bedding (fig. 17), in which B B represents the direction of the planes of bedding.

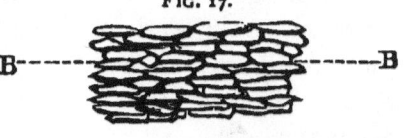

FIG. 17.

Fig. 18 indicates the change of direction which these minute and usually microscopic particles assume when the rock has undergone great lateral pressure. A fissile structure, called cleavage, is then set up in some direction other than that of the bedding planes B B; as, for example, in the direction C C, so that the rock splits more or less

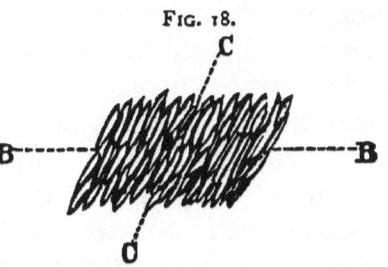

FIG. 18.

readily in that direction. This change in the minute structure of rocks which have been subjected to strong pressure was first demonstrated by Sorby, and fully explained by D. Sharp in the 'Quarterly Journal of the Geological Society,' and by Professor John Phillips in 'Report of the British Association, 1856.' It also affects any fossils which the rock may contain, squeezing and distorting them to a considerable extent. True schistose fission and slaty cleavage are seldom or never met with in rocks of eruptive origin, except sometimes in beds of volcanic ash, and occasionally in some of the older lavas, as shown by J. Arthur Phillips, neither does lamination occur in eruptive rocks, but a structure slightly resembling it is often to be noticed in sedimentary rocks which have undergone such great change (metamorphism) that they approach to, or are identical with, true eruptive rocks in their mineral constitution. This foliation consists in the segregation of any one mineral component of the rock along a more or less regular plane, and the result is a differentiation of the rock into a series of alternating layers of different mineral composition. These layers are often very thin, and at times scarcely to be discerned with the naked eye. Hornblende schist, for example, consists of alternating layers of hornblende and quartz; gneiss of layers of quartz, felspar, and mica. Gneiss may therefore be regarded lithologically as a foliated granite. Foliation has often been found to coincide with the original planes of bedding, as noticed by Ramsay, Darwin, Sterry-Hunt, and other observers, but this is not invariably the case. Metamorphic rocks form, as it were, a connecting link between the sedimentary and the eruptive classes, their pseudo-eruptive characters having been superinduced by the contact or proximity of highly heated eruptive matter. Thus, where basalts come in contact with limestones, the latter frequently become crystalline for some distance from the contact. Such metamorphism affects rocks sometimes on a small, sometimes on a large scale, occasionally influencing only a few inches, at other times extending for miles. It consists sometimes merely in physical, at others in chemical and physical changes, which frequently involve complicated atomic interchanges (chemical reactions), and symmetrical molecular rearrangements (crystallogenesis). By these means minerals are developed in a rock which it did not previously contain; and this process may take place without any accession of fresh elementary substances, analyses of the unaltered and the metamorphosed rock being sometimes nearly identical. The presence of hygrometric water, or quarry water, greatly facilitates such changes in the mineral constitution of rocks. This fact is very ably dwelt upon by Mr. John Arthur Phillips in several papers on the petrology of Cornwall published during the last few years in the 'Quarterly Journal of the Geological Society.' These matters will, however, be more fully discussed in the sections specially devoted to the changes which rocks undergo.

Although many admirable descriptions of volcanoes are to be found in most manuals of geology and in works specially devoted to the geology of volcanic districts,[1] yet it may be well to give here a brief description of the general structure of volcanic vents.

An active volcano may be defined as a passage or pipe which affords to deep-seated mineral matter, in a state of fusion, the means of transmission through the earth's crust, and of egress at its surface. A passive or extinct volcano is one in which this communication is obstructed, either by a plug of solidified lava, or by accumulations of fragmentary matter, a dissipation, temporary or permanent, of the eruptive energy, permitting the solidification of the molten matter. Should an augmentation of the eruptive force occur, the plug will either be shattered, and ejected in the form of lapilli

---

[1] The student may consult the works of Lyell, Scrope, Darwin, Daubeny, De la Beche, &c. with advantage; also some very interesting papers by Prof. J. W. Judd, entitled 'Contributions to the Study of Volcanoes,' published in the *Geological Magazine*, and the chapters devoted to this subject in *Geology for Students* by Prof. A. H. Green.

and ashes, or re-melted and poured out as lava, but, if it be unable to re-open the old passage, new vents may be produced, either within or without the lip of the crater.

Lava transmitted through a fissure or pipe and extruded at the surface may give rise to hills of a dome-shaped character.[1] Ashes and lapilli ejected from a vent become piled up around it, and in time form a conical hill on what may once have been a level surface. As, however, they are loose incoherent deposits, the hill will gradually acquire a slope at which they are no longer stable. On measuring the superficial inclination of hills composed of such ejectamenta, it has been found that the slope is usually about 30°.[2] If we pour sand upon a level surface, it forms a complete cone, but the loose volcanic materials come from below in the first instance, and since the pipe of the volcano is open, and ashes and lapilli are ejected from it and fall around it, the cone can have no apex. Some of the ejected matter, which is not carried away by the wind, showers down again upon the hill and around the orifice of the vent, but the law which governs the stability of these loose accumulations again prevents them from resting upon a very steep slope, and they are found to dip inwards towards the orifice as well as outwards down the slopes of the hill. The boundary line between these two slopes, which of course represents their greatest altitude, assumes a more or less annular form, and the inner slopes which dip towards the vent constitute a cup-like hollow, termed a 'crater.' Volcanoes, however, pour out lava as well as eject ashes, and these phenomena usually alternate. Lava in a viscous, pasty condition, rises through the pipe into the crater, where, after perhaps surging up and down for a time in a state of ebullition, it rises to the lip of the crater and runs over it down the sides of the hill and for some distance over the adjacent country. Sometimes

[1] As the Schlossberg at Teplitz, described by Dr. Ed. Reyer, Jahrb. k. k. geol. R. Anst. Vienna. 1879.
[2] See also paper by Prof. J. Milne in *Geol. Mag.* Decade II. vol. v. p. 337.

the mass of molten lava carries away one side of the crater, forming a great breach through which successive streams of lava are poured. The eruptions of lava may be succeeded by fresh ejections of lapilli and ashes, and these again may be followed by more lava streams, the hill eventually consisting of stratified fragmentary accumulations with interbedded flows of lava. Occasionally the lava is also forced through fissures in these deposits, forming dykes or wall-like masses which intersect them in various directions, usually, however, assuming a somewhat radiate disposition around the cone. When the volcano has done its work as a safety-valve the eruptions may cease for a time, and the vent may become plugged in the manner already described. Should a fresh eruption occur, it may force a new vent. Ashes are again showered out, lava is again poured forth, and a new cone is erected within the old one, or little cones and fumaroles are formed on the sides of the hill and dotted over the surrounding country.

CHAPTER V.

THE COLLECTING AND ARRANGEMENT OF ROCK SPECIMENS.

IN collecting specimens of rocks, it should be borne in mind that small pieces of compact and fine grained rocks answer the collector's purpose just as well as large ones, and often better, should he have but a limited space in which to store them. Small specimens are easier to get than larger ones, and in the course of a day's work, a much greater number of small specimens can be carried. In collecting for a museum, where there is plenty of available space, of course large specimens are best. About 5 inches by 4 inches will be found a convenient size when they are properly dressed, but if it should not be advisable to dress them in the field, larger pieces should be collected, as it frequently happens that a roughly broken block is reduced to half its original size before it is properly dressed. When a rock presents any large structural peculiarities, it will of course be necessary to collect proportionally large specimens in order to show the structure clearly. For ordinary private collections, specimens dressed to 4 in. by 3 in., or 3 in. by $2\frac{1}{2}$ in. are convenient sizes, or, if space be very limited, pieces about two inches square will suffice.

A hammer with a tolerably heavy head made of Swede-iron, with steel ends welded on and well tempered, but not so highly as to be brittle even when used on the hardest rocks, will be found to be best suited for collecting. The shaft should not be less than 13 or 14 inches in length; a tough wood such as ash answers very well for this purpose, and care should be taken in the selection of the wood. The eye into which the shaft is fitted ought not to be less than $1\frac{1}{8}$ inch in length by at least $\frac{1}{2}$ inch in breadth, and the head, which should have one end wedge-shaped, ought to be filed away slightly around the under opening of the eye to reduce the chance of breaking the shaft, as the fracture almost always takes place just under the head of the hammer. The author has been in the habit of using hammers with very heavy heads and with shafts long enough to serve as walking-sticks. Much heavier blows can be struck with a hammer of this kind than with a short-handled one, and their use does not necessitate such continual stooping. In some cases, however, a short-shafted hammer has its advantages, while for the purpose of dressing specimens one with a very long shaft is perfectly useless. When, therefore, a long-shafted hammer is taken into the field it is well also to carry a light dressing hammer. Short-shafted hammers are most easily carried in a small leathern frog with a flap, on the back of which are fixed one or two little vertical straps through which a waist belt is run; and this belt can also carry pouches for a compass and a clinometer. A strong canvas bag of tolerable capacity is necessary for carrying the specimens in, and it should have a little tab by which it can be loosely attached to a button on the back of the coat to prevent it from slinging forward when the wearer stoops. A good supply of paper should also be carried in which to wrap the specimens, and on the inside of each wrapper the precise locality from which the specimen is derived should be recorded.

These may seem trivial details, but neglect of them often causes disappointment and inconvenience. With regard to the best shape for the crushing end of the hammer-head, some prefer it flat and square, and others rounded. When it is slightly rounded the hammerer is less liable to be struck by splinters of stone, but for chipping purposes a flat square face is best, and the dressing hammer should always have one such termination.

In collecting, one of the first and most important things is to procure specimens which are unweathered or which have suffered as little as possible from atmospheric agency. Sometimes it so happens that a weathered surface of rock shows structural peculiarities which are especially worthy of note, owing to the different power which its component minerals possess of resisting disintegration and decomposition. Interesting specimens of this kind should always be collected so long as their transport will not lessen the number of more interesting unweathered specimens. It is only from the latter that a true knowledge of the normal mineral and chemical composition of rocks can be derived. The writer lays especial stress upon this, as he has at times been greatly troubled by being requested to determine rocks from badly selected specimens in an advanced stage of decomposition. Where quarries occur there is no excuse for collecting such rubbish. In other cases it is often a matter of difficulty to get unweathered samples, and sometimes any specimen is better than none ; still, as a rule, the collection of weathered chips is time wasted, and it is far better to take a little extra trouble in order to get good and typical pieces.

The specimens should ultimately be dressed with a small hammer, the piece of stone being held in the palm of the left hand, while with the right successive flakes and chips are struck off by sharp blows with the hammer. When very tough rocks are operated upon in this way it is by no means uncommon for the novice to end by getting a more or less rounded mass, covered all over with powdery, crushed surfaces, resulting from the bruises made by the hammer, and which do not show the character of the rock. With practice he will, however, soon ascertain the directions in which his blows will prove effective, and those where no amount of hammering will avail. When the specimen is properly dressed, one, or, still better, two little labels carrying numbers should be affixed to it. The use of two labels is desirable, since, if one becomes detached, the specimen can still be identified, so long as the other adheres. Ordinary strong gum answers fairly well, but in some cases, especially when the rock is soft and earthy, glue will be found preferable. Specimens are sometimes numbered with red sealing-wax varnish, but the figures are often difficult to find, and, when painted on rough surfaces, are not very legible. Of course, where numbers are used, the specimens must be carefully catalogued. In other cases labels an inch or more in length may be affixed, with the name and locality of the specimens written on them, but they have the disadvantage of covering a larger space than the little numbered tickets. Should labels be used, care should be taken to stick them on the worst dressed and least interesting parts of the specimens. When rocks are very soft and earthy, numerals may be scraped on them and form more permanent records than labels, which often become detached, even without handling. In arranging rock specimens, various systems of classification may be adopted. When they are intended to illustrate the petrology of any particular district or country they have merely a topographical arrangement which seldom admits of any really scientific classification, since eruptive and metamorphosed rocks have to be placed beside the sedimentary ones with which they are associated, and, so far as eruptive rocks are concerned, this does not always mean a chronological arrangement. The latter is certainly the right system to follow in dealing with sedimentary rocks, but with those of eruptive origin it is often very uncertain, and is of comparatively little value, at all events, in the present state of our knowledge; since eruptive rocks, almost identical in mineral composition, range from very early geological periods up to the most recent times. For the purpose of teaching petrology a classification based upon the mineral constitution of the specimens is doubtless the best, although, for general geology, the topographical arrangement is a useful one. When this is adopted in museums there should also be another small collection of rocks classified according to their mineral composition. A classification based upon structure is also to some extent to be commended, since it serves more or less as a grouping according to the conditions under which the rocks have been formed. Collections so arranged as to be illustrative of the stones used for building and ornamental purposes also have their advantages, but the former should always include weathered examples of the rocks to illustrate their powers of resisting disintegrating agencies, and, with this end in view, it would be most desirable to have an accompanying suite of weathered and partially decomposed stones taken from buildings, with the date when the building was erected recorded on them, or, if they have merely been used in the restoration of those edifices, the date of those repairs should be affixed to them, to show how much disintegration the stone has undergone since its surface was dressed and exposed in the building. In private collections, where rock specimens are usually arranged in cabinets, the drawers should not be less than two and a quarter inches in depth (inside measure), and deeper drawers are often very convenient. In museums, where space is ample, table-cases are best suited for the display of the specimens. Wall-cases are objectionable, because the specimens on the higher and lower shelves cannot be seen with any degree of comfort, while, if the rooms be badly illuminated, the wall-cases are almost certain to be worse lighted than any others. Glass cases standing away from the walls and fitted with shelves which range from the height of an ordinary table up to about five or six feet, and so arranged that the higher ones gradually recede more and more from the glass, are very good for purposes of study.

Under any circumstances the receptacles for the specimens, whether cases or drawers, should be well fitted so as to exclude dust as much as possible. For eruptive rocks to be properly displayed in museums they require quite as good or even better illumination than minerals. This is also desirable for sedimentary rocks, but is of less importance as a rule.

CHAPTER VI.

PRELIMINARY EXAMINATION OF ROCKS.

FOR the more general and preliminary examination of rocks the following implements will, if judiciously used (their use being backed by a moderate knowledge of mineralogy), be found sufficient for simple investigations. A stout-bladed penknife for testing hardness, &c. Small fragments of the minerals which constitute Mohs' Scale of Hardness. The diamond (No. 10 in this scale) may be omitted, as rock-forming minerals seldom have a hardness exceeding 8. A pocket magnifier, one of the ordinary pattern, having two or three lenses, will suffice for most purposes, but a good Coddington lens is also useful at times. There is, however, a disadvantage attending the use of these lenses when they are applied to the examination of rocks. This lies in the difficulty experienced by the observer when he attempts to examine the streak of minerals under the lens, especially when the minerals occur in very minute crystals or patches, as it is scarcely possible to hold a specimen, with a lens over it in focus, in one hand, and to work with a knife in the other. Laying the specimen on a table, and using a lens in one hand and a knife in the other, is a most unsatisfactory process, while the use of a lens fixed on an adjusting stand is scarcely better. To obviate this difficulty the author has devised a small lens with a clip, which can be worn on the nose like an eye-glass, and both hands are then at liberty—

the one to hold the specimen firmly, the other to use the knife or graver. This clip-lens is, moreover, better than a watch-maker's eye-glass, because it entails no muscular effort to keep it in place. It is better to have the lens mounted in a horn than in a metal rim, as it is less heavy, and consequently less liable to be accidentally shifted or displaced by the inclination of the head.[1] A bottle of hydrochloric acid with a glass rod attached to the stopper is useful in testing for carbonates. A small rough unglazed piece of porcelain may be advantageously employed for determining the colour of the streak afforded by minerals, but the same end may be attained by scraping some of the powder on a piece of white paper, and rubbing and griming this powder on to the paper with the side of a knife blade. The blade of the knife may be magnetised, and it then answers well enough for separating any substances attractable by a magnet from pulverised fragments of rock. In most cases, however, it is desirable to have a freely suspended or supported magnetic needle, since by its use in the examination of rocks containing magnetite, &c., repulsion as well as attraction may sometimes be observed. The operator may also, by holding a sheet of paper in front of the mouth and nose, prevent any undue disturbance of the needle by the breath. This is especially needful in examining rocks in which magnetite, &c., exists in only very small proportion. A small dressing-hammer, with a head about two and a half inches long, one end with a face about half an inch square and the other end chisel-shaped, and a little block of steel about two inches square and half an inch thick, and with one side polished, to serve as an anvil, will, together with a blowpipe, a small blowpipe lamp, some platinum wire, some pieces of well-burnt charcoal and a few fluxes and other blowpipe reagents, complete the list of apparatus needful for rough work; and by the skilful use of these simple appliances quite as much know-

---

[1] These lenses are manufactured by Messrs. Baker & Co., 244 High Holborn.

ledge may often be gained as by the employment of much more elaborate methods of investigation. It is, however, most desirable that the student should possess some knowledge of chemistry, the more the better, but for most purposes, in the determination of rocks, a fair knowledge of blowpipe analysis and some familiarity with the more common chemical reactions in the wet way will be found sufficient, especially when the observer is cautious in forming his conclusions, and bears in mind the old adage that 'a little knowledge is a dangerous thing.' Some acquaintance with the more common rock-forming minerals is also absolutely necessary. There are so many manuals and text-books which treat of these subjects that it is difficult to recommend any particular work, but a list of some of the most useful will be found at the beginning of this volume, and it may here be as well to remark that any attempt to study eruptive rocks without a fair knowledge of mineralogy may justly be likened to an attempt to read before the alphabet has been learnt. It is also very important that the student should have a good knowledge of physical geology, so that he may be enabled to make correct notes on the mode of occurrence of rocks and form sound deductions from his observations. In this part of the subject, reading, unless supplemented by some training in field work, will be found to be of comparatively little use. For those, however, who may not have had opportunities of getting any such training, the remarks embodied in the foregoing Chapters III. and IV., if carefully studied, may prove useful.

CHAPTER VII.

THE MICROSCOPE AND ITS ACCESSORIES.

IN describing the microscopes suitable for examining thin sections of minerals and rocks it will perhaps be best, in the interest of students generally, to begin with a description of the chief points to be attended to in the selection of microscopes of the ordinary patterns manufactured and sold in this country—instruments constructed to meet the requirements of the physiologist, the general microscopist, and the dabbler in science. Unfortunately no microscopes are yet manufactured in this country for the special study of micropetrology, but a good instrument of the ordinary type will be found to answer the purpose very well if a few comparatively inexpensive alterations be made in it. It is a somewhat difficult matter to recommend the microscopes of any particular maker. Those by Ross, Powell and Leland, Beck, Baker, Murray and Heath, Collins, Browning, Crouch, and Swift are all good according to the price. In the instruments by the three first-named makers the optical and mechanical arrangements are carried to the highest degree of perfection, while, in those by the others, there is a great amount of good workmanship in the more expensive microscopes, and, in some of the cheaper ones which they supply, the performance is very satisfactory. There are many other makers, besides those just mentioned, who turn out good instruments, but as the author has had little or no opportunity of testing them, he can only speak from experience, and, having principally worked with microscopes made by Baker, he thinks it only just to say that the performance of his higher class instruments is good, and that they would, if supplied with concentrically-rotating, graduated stages and a few other fittings, answer all the ordinary requirements of the petrologist. Mr. Watson of Pall Mall now manufactures microscope stands, specially suited for petrological work, at a moderate cost. (For description, see p. 307.)

Many observers prefer to work with binocular microscopes, but they seem to offer few advantages and are in some respects inconvenient, notably in the examination of objects under high powers, unless the prism be specially constructed for their use (as that by Powell and Leland).

They may be very advantageously employed in the examination of sections of vitreous rocks, of detached crystals, of pulverulent matter, as volcanic ashes and residues procured by levigation, but for the ordinary purposes of the petrologist a monocular instrument will in most cases be found to answer the purpose just as well, if not better.

It is *essential*—

(1) That the microscope should have a firm and tolerably heavy stand, and that the whole instrument should be free from vibration when the observer is focussing, or when he is moving the object.

(2) That the objectives should give a flat field, that they should be achromatic, and possess good definition and penetration. Better penetration is usually procured with objectives which have not a very wide angular aperture.

(3) That the stage be so arranged as to admit of the object being moved about an inch both from back to front and from side to side, and it is also essential that there should be an arrangement for rotating the object. If this rotation be concentric, so much the better. The stage should also be graduated so that the amount of rotation can be accurately measured.

(4) That the instrument be fitted with a polarising apparatus. The polariser should revolve with perfect freedom, and be so arranged that it can be removed, or, still better, displaced by turning on a hinge or pivot, so as to afford the means of instantly changing from polarised to ordinary illumination. The analyser should be fitted in a cap to slide easily over the eye-piece, so that it can be instantly removed, and there should be an arrangement so that the analyser and polariser can be accurately set with the longer diameters at right angles, or it may be

fixed in a sliding box or block just above the objective, as in the microscopes manufactured by Collins and one or two other makers.[1] On no account should the analyser screw into the top of the objective, as such an arrangement entails great trouble and loss of time.

(5) That the eye-piece of lowest power should be fitted with crossed cobwebs for centering and for goniometric measurements, &c., and with either a Beale's reflector, or, better, with a camera lucida (Wollaston's prism) for drawing objects. If the latter, care should be taken to see that it performs well, and that the whole field is properly illuminated. The drawing made by the first instrument is reversed; by the Wollaston's prism it is represented in its proper position.

(6) That the instrument be provided with a bull's-eye condenser, or with a silvered side-reflector, or a parabolic speculum, such as that devised by Mr. Sorby, and manufactured by Smith and Beck.

(7) That a stop be placed so that the body of the instrument can be kept in a horizontal position when drawings are being made with the camera.

The most generally useful objectives are the two-inch, the inch, and the half-inch, but higher powers are sometimes required. The objectives by some of the English makers are very perfect in their performance, but some of the Continental ones are exceedingly good, especially those by Hartnack and Gundlach, and they are less expensive than those made by many of the English opticians. A good objective should be perfectly achromatic (*i.e.* no fringes of colour should be visible around the object). It ought also

[1] A useful form of fitting for analyser and polariser, adapted for use with any microscope, has lately been devised by Messrs. Murray & Heath, of Jermyn Street.

to afford a flat field (*i.e.* when the object is moved from one part of the field to another it should appear well and sharply defined, the change in its situation not necessitating any alteration of focus).

Mechanical stages, such as those moved by racks and pinions, screws, &c., are not by any means necessary, although they are very convenient, especially when working with high powers. A well-fitted sliding object-carrier will be found to answer very well when the hands have had a little practice in slowly moving the object from back to front of the stage, while properly educated fingers are capable of giving a very steady transverse motion to the object. Spring clips, as usually fitted to the stages of the cheaper microscopes, have both advantages and disadvantages. Unless some kind of clip be used the object topples over when the microscope is placed in a horizontal position for drawing with the camera, or when the stage is rotated, while the ordinary little spring clips are nuisances when it becomes needful to examine sections which are in course of preparation, as they cannot accommodate thick slabs of plate glass, such as those to which the pieces of stone are cemented when being ground down. Sub-stages or supplementary stages (for carrying the polariser, spot lens, &c.), are of little or no use to the petrologist. They simply encumber the microscope with troublesome adjustments, which are seldom or never used, and which only get in the way. Moreover, when the polariser is fitted to a sub-stage a screen of some kind ought to be used to prevent the passage of extraneous light between the top of the polarising prism and the bottom of the stage, otherwise erroneous conclusions may be arrived at with regard to singly refracting substances. It is far better that the polariser should be directly attached to the under surface of the stage.

An achromatic condenser is occasionally useful to the petrologist when examining feebly translucent sections. If the stage of the microscope be moved with rackwork a Maltwood's finder will be found convenient.

This consists of a series of minute squares photographed upon a glass slide, each square bearing two numbers (fig. 19). This finder affords a ready means of registering the exact position of any minute object in a preparation. Unless a finder be employed, much time is often lost in the endeavour to re-discover any very small object or any particular portion of a section which has been previously observed. It is used in the following manner:—

FIG. 19.

$\frac{1}{1}$	$\frac{2}{1}$	$\frac{3}{1}$	$\frac{4}{1}$	$\frac{5}{1}$
$\frac{1}{2}$	$\frac{2}{2}$	$\frac{3}{2}$	$\frac{4}{2}$	$\frac{5}{2}$
$\frac{1}{3}$	$\frac{2}{3}$	$\frac{3}{3}$	$\frac{4}{3}$	$\frac{5}{3}$
$\frac{1}{4}$	$\frac{2}{4}$	$\frac{3}{4}$	$\frac{4}{4}$	$\frac{5}{4}$

A small peg or stop is screwed into the object-carrier so that one end of an ordinary slide may abut against it.

When any particular object or spot is noticed, and it is desirable to record its position, the object is removed, without touching any of the mechanical adjustments of the stage. The finder is then placed on the stage, with one end closely in contact with the stop against which the end of the object-slide previously rested; the instrument is re-focussed if necessary, in order to read the numbers in the centre of the field. These numbers may then be recorded either on the label of the slide itself or in a note-book. When it is again requisite to find the object, the finder is put on the stage as before, and is moved about until the recorded numbers are in the centre of the field. It is then removed, the preparation is put in its place, and the required object will be found to occupy the place of the recorded numbers.

Another finder, applicable to microscopes which have not mechanical stages, consists in a metallic arm, hinged or pivoted to a fixed portion of the microscope above the stage. The arm is curved, and terminates in a point. When it is needful to register the exact position of some object in the field of view, the point of the arm is daubed with ink, and it is then brought down upon a paper label pasted on one end of the slide which carries the preparation. There it imprints a dot. By again bringing this dot under the point of the arm the desired object can at once be brought into the field. To those who devote much time to making microscopic drawings it will be useful to have a small movable needle or indicator N (fig. 20), movable by a milled head H, fitted into one of the eye-pieces. By this means it is easy to find the spot at which the draughtsman was last looking.

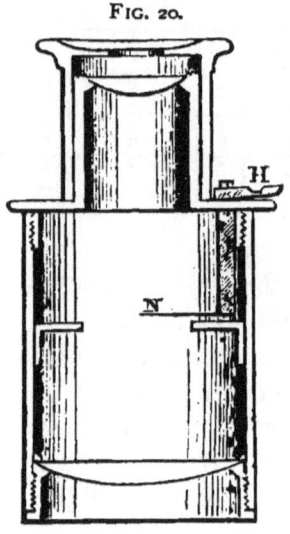

FIG. 20.

Brook's double nose-pieces, or the triple or quadruple nose-pieces devised by various makers, are convenient, inasmuch as they facilitate the rapid interchange of objectives of varying power. Their workmanship is, however, seldom sufficiently good to render their use desirable with high powers, and their constant employment does not tend to improve the performance of the fine adjustment, with the tube of which they are connected. Those nose-pieces which are turned up, so that the objectives do not stand parallel when screwed on, are to be preferred, as they permit the free use of objectives of varying length and of different foci, without the risk of jamming an objective against the stage, and so deranging the mechanism of the nose-piece and fine adjustment, together with other possible damage to the instrument.

In selecting microscopes, those on the new Jackson model are greatly to be preferred for rigidity, steadiness, and accuracy of centering.

The use of very deep eye-pieces should be avoided; an A and B or a No. 1 and No. 2 eye-piece will, as a rule, be all that are requisite. Deep eye-pieces entail feeble illumination, and augment any error which the objective may give rise to.

Frog-plates, live-boxes, compressoria, and other like apparatus, usually supplied with microscopes, are of no use whatever to the petrologist.

It is occasionally very important to be able to determine with some degree of precision the angles presented by sections of crystals as they occur in sections of rocks, or the angles of minute individual and unattached crystals. In the latter case it is essential that the crystal should be so arranged that the faces whose mutual inclinations have to be measured are situated in planes parallel to the axis of vision. In the sections of crystals which occur in thin slices of rocks, it is frequently a matter of considerable difficulty to ascertain in what direction the section has been cut; and such sections are not very often cut in directions precisely at right angles to, or parallel to, the principal crystallographic axis, or to any one of the lateral axes; consequently measurements of the angles offered by such sections are, as a rule, only approximate. Although, therefore, a very exact goniometer is desirable, a tolerably good one will usually suffice. The simplest method of measuring angles under the microscope is to place the microscope in a horizontal position, and by means of the camera to draw the outlines of the two faces to be measured on a good-sized sheet of paper; these lines should then be prolonged, and the angle at which they lie to one another can easily be measured with a common protractor. By this means, however, measurements cannot be got to within less than 30' or at best 15'. Still, if the section be not cut in exactly the proper plane, the results given by a more precise instrument would not be of any greater practical value. A very good goniometer, that known as Schmidt's, a form of which is manufactured by Ross and Powell and Leland, consists of a positive eye-piece in which a cobweb is placed; around this eye-piece there is a graduated brass circle about four inches in diameter, and an arm carrying a vernier and set-screw is attached directly to the eye-piece, so that when the arm is moved the eye-piece itself turns together with the cobweb. In using this instrument, the cobweb is aligned on one of the faces of the crystal, and the movable index is brought to O on the circle.

The arm is then turned until the cobweb is aligned on the adjacent face, the set-screw is turned to clamp the arm, and the number of degrees, minutes, &c., through which the arm has travelled is read off by means of the vernier, over which a small magnifier is usually placed. When the angles of crystals occurring in sections of rock which are not very translucent have to be measured by this instrument, difficulty is often experienced in seeing the cobweb distinctly, and this is one of the most serious drawbacks to the use of this kind of goniometer for petrological purposes. Its utility would probably be increased if one half of the field were obscured by the insertion of a blackened semicircle of metal within the focus of the eye-piece instead of the cobweb. Measurements of the angles of crystals may also be made by the use of an eye-piece carrying crossed cobwebs, and a concentrically revolving stage, if the stage be graduated on the margin; but as such stages are only furnished with first-class instruments, the majority of students are not likely to avail themselves of this method.

A microscope specially constructed for mineralogical and petrological purposes has recently been devised by Prof. Rosenbusch, of Strassburg. The chief advantages of this instrument consist—

(1) In the facilities for turning an object in its own plane between fixed crossed Nicols, the rotation being concentric with the axis of vision.

(2) In the ability to read off accurately the angle through which the object may be turned in a horizontal plane by means of the graduation around the circular stage.

(3) In the facility with which the polariser and analyser can be displaced and replaced, and the means by which the exact position of the principal sections of the polariser and analyser can be noted.

(4) Where the total extinction of light by means of crossed Nicols interferes with the researches on

any mineral, means are provided for facilitating observation under such circumstances.

The peculiarities in the construction of this microscope consist in the tube which carries the eye-piece and objective *m n* (fig. 21), being, as it were, suspended within an outer tube *o p*, its only attachment being at the top at *b c*. A block, *r*, is fixed between the inner and outer tubes to prevent any rotation during focal adjustment. This coarse adjustment is effected by hand, as in some of the cheaper English microscopes, the thumb and forefinger sliding the inner tube up and down by pressure on the disc *d e*, other fingers being applied to the top, *b c*, of the fixed tube. The fine adjustment consists of a micrometer screw, shown at *a*. The unattached portion of the inner tube is steadied in the outer one by means of a spring and three little screws set horizontally and capped with scraps of parchment. The arm of the microscope carries two screws with milled heads, one of which is shown at *h*. These are set at right angles to one another, and serve to centre the tube in a manner presently to be described. Each eye-piece carries two cobwebs within it, which intersect at right angles in the centre of the field. To the outside tube of each eye-piece a small peg is fixed, which slides into a corresponding slot in the top of the inner movable tube of the microscope. This arrangement prevents any rotation of the eye-pieces, and so keeps the cobwebs in a fixed posi-

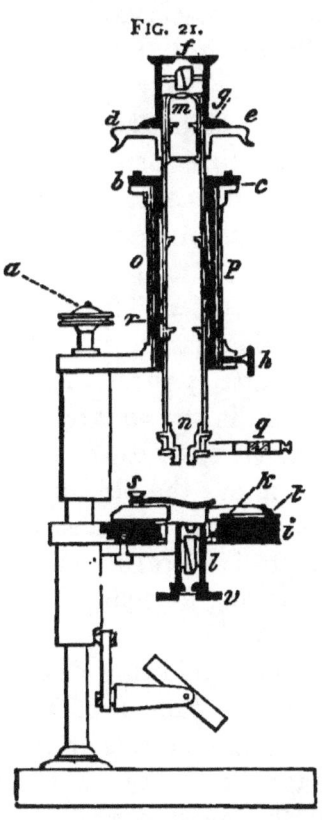

FIG. 21.

tion. An analyser, *f*, fitting in a brass cap, slides easily over the top of the eye-piece. The bottom of the cap is surrounded by a bevelled flange, *g*, which is graduated to 5°. An index mark on the plate, *d e*, serves to record the angle through which the prism may be rotated. The stage, *i*, of the microscope has a circular form, and a circular plate or object-table, *k*, is arranged so as to revolve horizontally on it. This table is graduated on its margin, and an index to record the amount of the revolution which may be imparted to it is attached to the front of the fixed stage at *t*. Beneath the stage is set an easily displaceable polariser, consisting of a Nicol's prism, which revolves within its external tube by means of the disc, *v*, which is graduated to 10°, and has its index marked on the fixed outer tube, *z*. This polariser does not turn when the object-table is rotated, but remains unaltered in position. A plate of quartz, 3·75 millimeters thick and mounted in a little brass fitting, is shown at *q*. It slides into a corresponding slot, situated close to the lower end of the inner microscope tube and above the objective, which is omitted in the figure. A small plate of calcspar for making stauroscopic measurements is also supplied with the microscope, together with a brass ring for fixing it above the eye-piece and beneath the analyser.

The following directions for using this microscope are extracted from Professor Rosenbusch's paper.[1]

If any particular spot in an object, such as a granule of magnetite, be brought exactly under the point of intersection of the eye-piece-cobwebs, *i.e.* in the middle of the field of vision, and the object-table be then turned in its horizontal plane, the inner tube of the microscope will be found to hang neither vertically nor concentrically without the intervention of the centering screws, while the spot under observation will not remain in the *centre* of the field, and

---

[1] *Ein neues Mikroskop für mineralogische und petrographische Untersuchungen.* H. Rosenbusch. Neues Jahrbuch für Mineralogie. 1876.

under the point of intersection of the cross-bars $a\,a$ and $\beta\beta$, (fig. 22), but will describe an eccentric circle somewhat in the manner shown in fig. 22. The tube of the microscope must therefore be placed vertically; in other words, the instrument must be centered by means of the two centering screws, one of which is shown at h in fig. 21. It will be seen by fig. 22 that the optical axis of the microscope is at o_1 and not at o, and, in order to get proper centricity in the movement, the spot o_1 should be made to coincide with o. To effect this the end of the tube must be moved in the direction of o, o_1. By means of one centering screw it will be driven in the direction o, ν, and by the other in the direction o, μ. When these adjustments have been properly made, the spot should be brought exactly under the intersection of the cross-bars in the eye-piece, and should remain there during the revolution of the object-table.

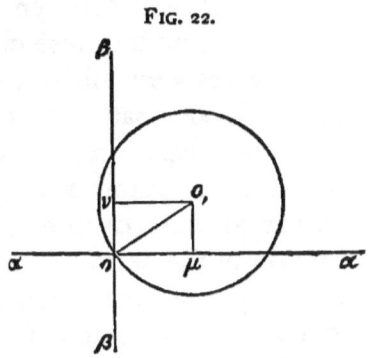

FIG. 22.

When this operation is once properly performed, any other spot or part of an object which may be brought into the field will, upon rotation of the object-table, be found to revolve concentrically, so long as the same eye-piece and the same objective are used, but if one or other of them be changed, it will usually be necessary to re-centre the instrument. This will, however, generally entail only a slight alteration of the centering screws. The movement imparted to the microscope tube by these screws tends to throw the analyser slightly out of position with regard to the polariser, but the inventor finds that this produces scarcely any appreciable error. In testing the pleochroism of a mineral, the object-table bearing the section may be revolved above the fixed polariser, or the polariser may be turned beneath the stage, the graduations affording facilities for determining the position of the pleochroitic maxima.

The principal directions of vibration in a mineral section may be determined by inserting it in the maximum extinction of light between crossed Nicols, but since the eye is incapable, under these circumstances, of appreciating in certain cases very slight differences in the transmission of light by depolarisation, a calcspar plate is inserted in the stauroscope between the analyser and the section of the mineral under examination. The interference figure of the calcspar appears distorted, until a direction of principal vibration in the section coincides with that of the polariser. In stauroscopic measurements very precise results may be arrived at by the employment, not of ordinary white light, but of monochromatic light derived from a coloured gas flame; this method, however, although useful when an ordinary stauroscope is used, is inapplicable to microscopic research. In microscopic examinations a plate of quartz $3^.75$ millimeters thick is used instead of a calcspar plate; where this is employed a monochromatic field is procured. When the principal direction of vibration in the analyser is turned at a different angle to that of the polariser, the field will become changed to various colours, where doubly refracting bodies are situated in the field of view, and their principal directions of vibration do not coincide with that of the polariser. By turning the object-table until such coincidence is arrived at, a purely monochromatic field will be produced; very slight movement of the object will again suffice to destroy the monochrome. The employment of such a quartz-plate is most useful when very feebly double-refracting media are being examined, and also for detecting isotropic particles in rocks with admixtures of amorphous paste of a doubtful character. In addition to the apparatus here described this microscope as manufactured by Fuess of Berlin (Alte Jacob Strasse, 108) is supplied with three eye-pieces and three objectives of Hartnack's make, which give a range of amplification varying from about 90 to 1,150 diameters, and by the use of the different eye-pieces affording a series of 9 different powers. An eye-piece micrometer and an apparatus for heating objects under examination, and recording the temperature by means of a thermometer, are also supplied with this instrument. Some useful modifications of Prof. Rosenbusch's microscope have been made by Prof. Rénard and described by him in the 'Bulletins de la Société belge de Microscopie,' tome iv. 1877-78. In the 'Neues Jahrbuch für Mineralogie, &c.,' 1878, p. 377, Professor A. von Lasaulx has described methods of converting ordinary microscopes so that they can be employed for the examination of minerals in convergent polarised light. Another paper by the same author, *op. cit.* p. 509, written a couple of months later, describes the construction of a polariscope, suitable for purposes of demonstration, which consists in part of the tube and Nicol's prisms of an ordinary Hartnack's microscope. A microscope devised for mineralogical and chemical purposes was devised some years since by Dr. Leeson. It was manufactured and improved by Mr. Highley, by whom it is described in the 'Quarterly Journal of Microscopical Science,' p. 281. Another microscope, specially constructed for the examination of substances in hot acid solutions and corrosive fluids, has been devised by Dr. Lawrence Smith. In this instrument the stage is placed above the objective, and the object is viewed from the under surface of the slide.[1]

CHAPTER VIII.

METHOD OF PREPARING MINERALS AND ROCKS FOR MICROSCOPIC EXAMINATION.

THE preparation of thin sections of minerals and rocks for microscopic examination, although effected by simple means,

[1] *American Journal of Science*, 2nd series, vol. xiv. 1852. The instrument is figured in *How to Work with the Microscope*, by Dr. Lionel S. Beale.

presents numerous difficulties to those who have had no previous experience in work of this kind. The object of the present chapter is to supply plain instructions concerning the needful appliances and the methods of manipulation by which such sections may be successfully made.

It is true that sections may be prepared by lapidaries,[1] and that the student is thus spared considerable labour and loss of time; but he will find, at all events in the earlier stages of his work, that there are certain advantages which he will derive from the preparation of his own sections. These advantages consist mainly in the facilities which he will have for testing the hardness of minerals, their deportment with chemical reagents, and the different appearances which they present when examined at intervals while the process of grinding them thinner and thinner is being carried on. The apparatus needful for such work is of a very simple kind, but more or less complex appliances for cutting and grinding will be found advantageous.

As it is desirable to lessen the labour of grinding as much as possible, the first thing to be done is to procure a thin chip or a thin slice of the mineral or rock about to be examined. A square inch is a convenient size for the chip or slice, as such a piece will often undergo considerable diminution before it is reduced to a sufficiently thin state. Chips may be procured by using a small hammer, but frequently a number of flakes have to be struck off before one of suitable size, thinness, and flatness is got. When the specimen is very small, and difficult to hold in the hand while the hammer is used, a satisfactory chip may often be procured by holding the fragment in a suitable position on the edge of a cold chisel either let into a block of wood (fig. 23), or screwed into a vice, but then the

FIG. 23.

[1] Mr. F. G. Cuttell (52 New Compton Street, Soho) prepares admirable sections.

operator must take care of his fingers. In the chipping of very hard rocks it is also advisable to protect the eyes, especially when the hammerer is not well practised in stone-breaking. For this purpose a pair of wire-gauze spectacles will be found useful. When cleavable minerals are to be dealt with it is best to avail oneself of the cleavage, but also to note in which direction of cleavage the plate is struck off, and, if it be desirable to make a section in some other plane than that of cleavage, a slitting or sawing process, hereafter to be described, is the only way in which such a section can be procured.

When a suitable chip has been struck off the specimen, the first thing to be done is to grind one side of it perfectly flat. This may be accomplished either by grinding it by hand on a flat cast-iron plate with moderately fine emery and water, or by using a machine with a revolving leaden lap, similarly charged for the purpose. The former method is the more tedious, and, although preferred by some people, is far less convenient than the latter, supposing the operator to have a suitable machine at his command. There are various forms of machines which have been devised for this purpose, some of them being worked by a treadle and others by hand; the latter are the more portable, but the former are usually considered easier to work. Machines of both kinds are manufactured by Füess, of Berlin, and other makers. Good treadle machines, devised by Mr. J. B. Jordan, of the 'Mining Record' Office, may be procured from Messrs. Cotton & Johnson, 21, Grafton Street, Soho, and will be found to be well suited for the purpose. These machines are supplied with slitting discs for sawing off thin slices of rocks or minerals, and with laps for grinding them down to the requisite degree of thinness.

The following details of construction, extracted from the Journal of the Quekett Microscopical Club, together with the use of the illustration, have been kindly furnished by the inventor :—'As will be seen from the diagram below, this machine consists of a wooden frame-work, *a a*, supporting a crank-axle and driving-wheel, the latter being two feet in diameter; the top part of this frame is formed of two cross-pieces, *a'*, fixed about an inch apart, as in the bed of an ordinary turning-lathe; into the slot between them is placed a casting, B, carrying the bracket for the angle-pulleys, C; this casting is bored to receive the spindle, D,

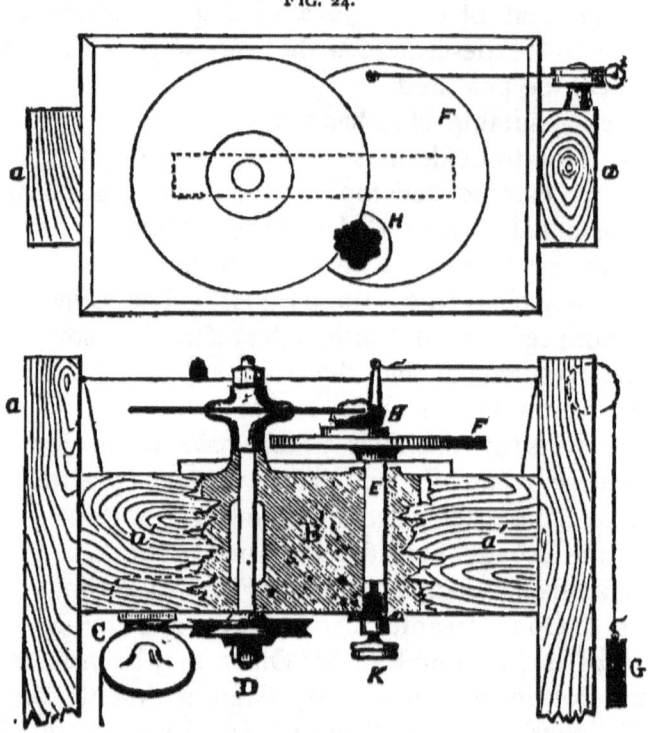

FIG. 24.

which, by means of a treadle, can be made to revolve at the rate of 400 or 500 revolutions per minute; it is also fitted with another spindle, E, having a metal plate, F, fixed on the top, for carrying the small cup, H, to which the specimen is attached by means of prepared wax. This method of mechanically applying the work to the slicer is far preferable to holding it in the hand in the ordinary way;

Preparation of Sections.

the requisite pressure against the cutting disc is regulated by the weight, G, and the thickness of the slice by the thumb screw, K, on which the spindle rests. By this means, it is possible to cut thin and parallel slices—the thinness of course varying according to the strength of the rock which is being operated upon. The slitting disc is made of soft iron, eight inches in diameter, and about $\frac{1}{50}$ of an inch in thickness, and it is fixed on the spindle, D, between two brass plates or washers, four inches in diameter, by means of the nut, n. The lap or grinding disc is eight inches in diameter, of lead or cast iron about $\frac{3}{8}$ of an inch thick in the centre, and having rounded edges and slightly convex sides; this form facilitates the grinding of uniform thinness, there being always a tendency on a flat surface (which soon wears hollow) for the edges of the section to grind away before it is sufficiently thin towards the centre.'

In using such a machine for slitting off slices the edge of one of the thin iron discs should be charged with diamond dust. This should be worked into a paste on a slab or in a small watch-glass (a stand for which may be made with the rim of a pill-box cover), with a little sweet oil, and the mixture taken up in small quantities on the end of a crow-quill suitably cut, and it should then be applied carefully to the edge only of the disc, the disc being slowly turned by hand for a short distance, say an inch or two, and afterwards rubbed in hard with a short but thick piece

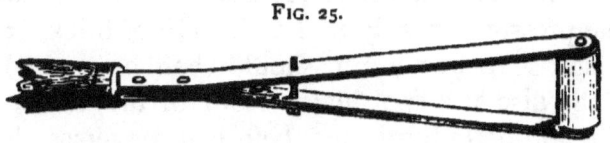

FIG. 25.

of glass cylinder about half an inch long, fitted to a pronged handle in such a manner that it acts as a roller (fig. 25). This process should be continued until the entire edge of the disc has been well charged. The piece of stone to be slit should then be fixed firmly in the small metal cup which is afterwards to be clamped in the movable arm or plate provided for its reception. The chip or fragment of stone may be fixed in the cup by means of Waller's wax, otherwise known as red cement. Small fragments of the cement should be placed in the cup and the whole held over a spirit lamp or a Bunsen's gas-burner until the wax is fairly melted. The piece of stone, previously warmed, is then pressed firmly into the wax.

It is well to press the cement while yet warm closely round the fragment, which is best done with a cold metal point. It may then be allowed to remain until it is quite cold. After fixing the cup into its arm or plate, the latter should be adjusted to the proper height for the disc to make the first slice. Suitable pressure should then be applied either by hand, by a pulley and weight, or by an elastic spring, fitted by one end to an upright rod on the table of the machine and by the other to a stud fixed on the carrying arm. Under any circumstances it is better to assist the regulation of pressure by hand. It should also be observed that, in commencing the process of slitting, the edge of the disc should first be brought into contact with a comparatively flat or rounded surface of the fragment of stone and not with a sharp edge, as, in the latter case, the diamond dust will probably be stripped off during the first revolution of the disc. This is a point to be very carefully observed: indeed it is better for the first few revolutions to apply the pressure entirely by hand. Oil-of-brick, or a mixture of soft-soap and water, should then be applied to the edge of the disc at a spot just in front of the stone, so that it may be properly lubricated before it traverses the hard surface, and on no account ought the disc to become dry while cutting, or the diamond edge will instantly be lost. The application of the lubricant may be made either by means of a brush held in the left hand or by a dripping apparatus, such as a tin pot with a very small tap fixed in the bottom. The disc should next be set in motion and a steady pressure and constant lubrication kept up until the slice is cut off. The carrying arm should then be raised, say one-eighth or one-sixteenth of an inch, according to the tenacity of the mineral or rock which is being dealt with, and a like process should be repeated until the second cut is finished, when the slice is ready for grinding. The processes connected with the grinding of a sawn-off slice and of a hammer-chipped fragment are identical. The leaden lap should be substituted for the slitting disc. Two pots or saucers should be at hand, the one filled with moderately fine emery and the other with water. A house-painter's brush (a small sash tool as it is technically termed) should then be dipped in the water and afterwards in the emery, and the resulting paste smeared all over the upper surface of the leaden grinding-lap. The machine should then be set in motion and the slice or fragment be firmly pressed on the surface of the lap by the fingers of one or of both hands, care being taken to keep the finger-tips clear of the revolving lap. A little practice will soon teach the operator the best way of doing this. When a good, even surface is procured in this way, the slice or chip ought to be carefully washed and wiped to free it from all adhering particles of coarse emery, and then the somewhat rough surface should be rendered as smooth as possible by grinding the fragment by hand on a flat brass slab, or on a slab of thick plate glass, about six or seven inches by four or five inches in diameter, smeared with the finest flour-emery and water. The motion of the hand in grinding should be a circular one, and it should be carried systematically all over the plate, so that the latter may not become unequally worn. When a perfectly smooth surface is procured the process must be stopped and the fragment again thoroughly washed and cleansed from all adhering emery. The next process consists in cementing the smooth surface of the stone to a small slab of plate glass about two inches square and about a quarter of an inch or more in

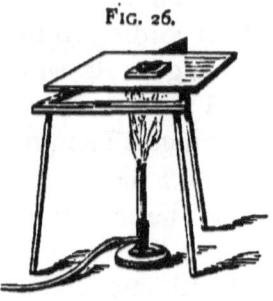

Fig. 26.

thickness, the edges being previously ground roughly on the lap to avoid the risk of cutting the fingers, in case it should slip when pressed on the revolving disc." One of these small glass slabs should be placed upon an iron, brass, or copper plate, supported either on a tripod or by other means, over a Bunsen's gas jet (fig. 26) or a spirit lamp,[1] and a few scraps of the oldest and driest Canada balsam which can be procured should be laid upon

Fig. 27.

the top of the glass, the piece of stone to be cemented also being laid on the iron plate (but not on the glass slab) with its smoothly ground surface uppermost. The jet or lamp should now be lighted and the gradual liquefaction of the balsam carefully

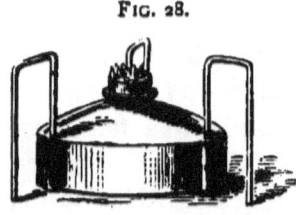

Fig. 28.

watched. As soon as the balsam liquefies (it ought on no account to be allowed to approach ebullition) the piece of stone should be taken up, reversed, and its smooth surface pressed into the balsam and on to the surface of the glass slab (fig. 27). The slab should then be pushed to the edge of the hot plate so

---

[1] A good form of spirit lamp (fig. 28) made for cooking purposes, but admirably adapted for microscopical mounting, is sold by Scott & Son, 42 Bedford Street, Strand, London, and costs less than two shillings. It is made of tinned iron; the wick is stuffed into a tinned iron cylinder an inch in diameter, and a cap with a brass collar screws over the burner when it is not in use. On the sides of the lamp are three small sockets which carry bent iron wires, thus forming a strong tripod upon which a copper or iron plate can be placed for mounting.

Preparation of Sections. 67

that it can be conveniently removed by a small pair of tongs. The tongs best adapted for this purpose consist of an ordinary wine-cork cut in half, the separate halves being fixed to a jointed arm, such as an old pair of compasses or dividers (fig. 29). With these the corner of the slab should be firmly held, and it ought then to be placed upon a piece of wood or thick pasteboard laid on the table and the piece of stone firmly squeezed and held down on the glass slab until the balsam begins to harden. For this purpose the cork ends of the tongs may be used, as the stone is usually too hot to be fingered. All these processes must be rapidly performed. The slab should then be taken up and examined from its under side to see that no air-bubbles have been included between the glass and the stone.

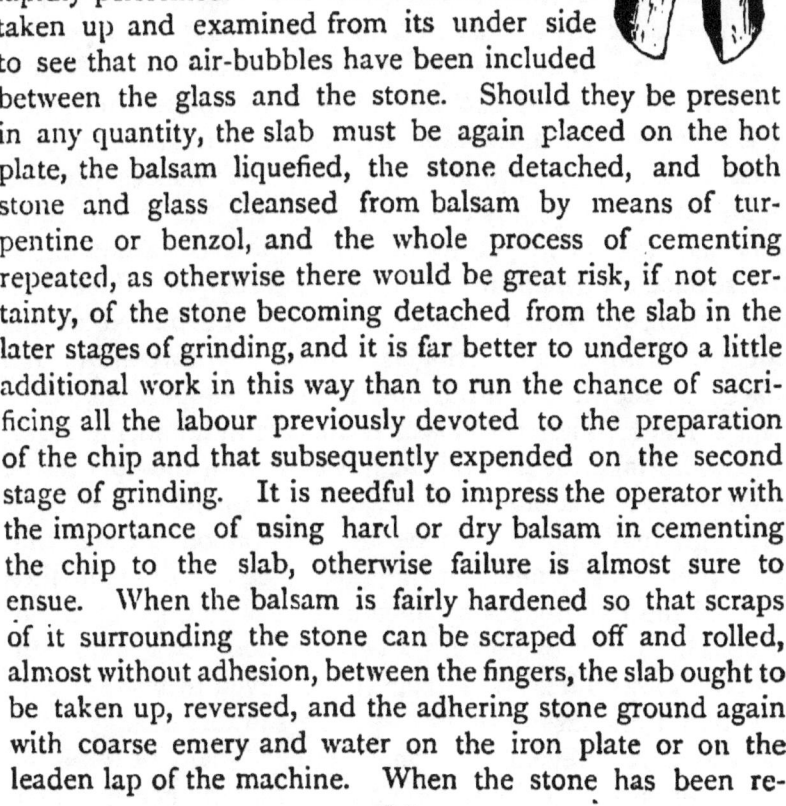

FIG. 29.

Should they be present in any quantity, the slab must be again placed on the hot plate, the balsam liquefied, the stone detached, and both stone and glass cleansed from balsam by means of turpentine or benzol, and the whole process of cementing repeated, as otherwise there would be great risk, if not certainty, of the stone becoming detached from the slab in the later stages of grinding, and it is far better to undergo a little additional work in this way than to run the chance of sacrificing all the labour previously devoted to the preparation of the chip and that subsequently expended on the second stage of grinding. It is needful to impress the operator with the importance of using hard or dry balsam in cementing the chip to the slab, otherwise failure is almost sure to ensue. When the balsam is fairly hardened so that scraps of it surrounding the stone can be scraped off and rolled, almost without adhesion, between the fingers, the slab ought to be taken up, reversed, and the adhering stone ground again with coarse emery and water on the iron plate or on the leaden lap of the machine. When the stone has been reduced sufficiently to transmit light, great care must be taken, and, if the section be very thin (*i.e.* if the stone be naturally rather opaque in thin plates), the pressure upon it should be diminished. The process of grinding with coarse emery must not be carried too far, as, when the section is extremely thin, it may often be entirely removed by one or two turns of the lap. The slab and its adhering section should then be thoroughly washed and freed from all traces of emery, and the final grinding conducted by hand on the brass plate or glass slab with flour-emery and water. In the very latest stages a few drops of paraffin may advantageously be used in order to diminish friction. During this final grinding the section should be frequently examined under the microscope, but must be thoroughly washed and cleansed prior to each examination and a drop of turpentine placed on its surface to increase its transparency. Now is the time to apply tests to the different component minerals, if the section be a rock or an impure mineral, and doubt exist as to the nature of any of the substances present. Some operators employ rouge for polishing the section in its very last stage of preparation, using a piece of parchment as the surface on which to polish it. The slab, with the adhering section uppermost, must once again be placed on the hot plate, while a watch-glass containing turpentine and placed on the rim of a pill-box or other support should be near at hand. When the balsam is thoroughly liquid, the operator should take the slab off the hot plate with the cork tongs, and by means of a blunt-ended wire (fig. 30), held in the right hand, gradually push or slide the section off the slab into the turpentine contained in the watch-glass. He ought then to hold the watch-glass by means of a wire ring, or a pair of crucible-tongs or forceps, over the lamp or gas jet sufficiently long to heat the turpentine or even to make it boil, but the watch-glass must be kept at a suitable distance from the flame to prevent ignition. It may then be replaced on its support, and the section should be very gently turned

Preparation of Sections.

over and washed in the turpentine by means of one, or two, small camel's-hair brushes. A glass slide (those in ordinary use for microscopic preparations are perhaps as good as any for size; 1 in. by 3 in.) must then be placed on the hot plate, which, however, should have been allowed to cool beforehand; a drop of ordinary fluid Canada balsam placed

FIG. 30.

carefully on the middle of the slide and the gas or lamp lighted underneath. The section should then be taken out of the turpentine by means of a needle mounted in a handle (fig. 31). This may be done by gently pushing the needle underneath

FIG. 31.

the section and slowly raising it, when the section will usually adhere to the side of the needle. Superfluous turpentine may be removed from the section by touching its lower edge with the finger, on to which the turpentine will flow. During this short

FIG. 32.

process the balsam on the slide ought to be closely watched to see that it does not get overheated. Should it boil it is useless to proceed until another slide with fresh balsam has been warmed sufficiently, without boiling. The lower edge of the section should then be placed on the balsam, the upper edge being supported by the needle, and it should be allowed to subside gently on the balsam, so that no air-bubbles are included. The slide must be at once removed from the hot plate, or, better still, it should be removed before laying the section upon it. The section can be gently moved about by means of a needle, to induce the balsam to pass over its edges, and the needle may then be used horizontally to drag the balsam over the upper surface of the section. If this cannot be easily accomplished, an additional drop of balsam may be placed on the top of the section, and the whole very slightly heated. When the section is completely covered with balsam, a thin and clean glass cover of suitable size should be taken in a pair of forceps, held for an instant over the gas or lamp flame, and be let down gradually on the section. It should be gently pressed on the surface to bring it close to the section and to drive out any air-bubbles; a slightly rotatory motion will be found useful for the latter purpose.

The slide may then have most of the superfluous balsam scraped off, and be numbered or labelled by means of a writing-diamond, and set aside to dry. When what remains of the superfluous balsam is found to be tolerably hard, it should be removed as much as possible by a knife with a square end, fig. 32 (if one with a point be used there is danger of dislodging the cover), and the remainder may be cleaned off with a rag dipped in turpentine, or, better still, in benzol. The preparation is then complete, unless the operator likes to stick paper labels on the ends of the slide. The writing on paper labels has the advantage of being more easily legible than diamond-writing, but brief names or numbers should always be first scratched on the glass with a diamond, as the scratches help the paper label to adhere more firmly, and, what is still more important, they are *permanent* records, so that if labels become detached the sections can still be identified. Some mounters etch the names and numbers on the ends of their slides by hydrofluoric acid, but the marks left by a writing-diamond are more legible and take far less time to do. In preparing sections of very soft and friable rocks the following process, communicated to me by Mr. John Arthur Phillips, may be had recourse to :—

The chip, which may to some extent be previously hardened by saturation in a mixture of balsam and benzol until thoroughly impregnated with it, and afterwards dried, should be gently ground, or filed down until a smooth, even surface is procured; this surface must then be attached to a piece of a glass slide cut about an inch square, and this again fixed in a similar manner by old balsam to a thicker piece of glass if needful, so that it can be conveniently held whilst the grinding is carried on. When it is reduced to such a degree of tenuity that it will bear no more grinding, even with the finest materials, such as jeweller's rouge, and when the removal of the section from the glass to which it is attached would almost inevitably result in the destruction of the preparation, the lower piece of glass should be warmed and separated from the upper piece which bears the section, and this, with its attached section, should be again cemented by the under side of the glass to an ordinary glass slip, covered in the usual way, and, if the edges of the section, or its glass, be disfigured by grinding, a ring or square margin of Brunswick-black, or asphalt, may be painted over the unsightly part.

Mr. E. T. Newton, the Assistant Naturalist of the Geological Survey, who has successfully examined the microscopic structure of many varieties of coal, has favoured me with the following notes upon the methods employed by him in making his preparations:—

'One important point to be noticed at the outset is, that nothing like *emery* powder can be used for the grinding, as the grains embed themselves in the softer substance of the coal, and, when the section is finished, will be seen as minute bright spots, thus giving to the section a deceptive appearance. For the rough grinding an ordinary grindstone may be used, and for the finer work and finishing, a strip of "pumice stone" (or corundum stick), and a German hone (or water of Ayr stone). The form of these which has been found most convenient is a strip about $1\frac{1}{2}$ inches wide and about 6 inches long; the thickness is immaterial: one of the broader surfaces of these must be perfectly flat. Having selected a piece of coal with as few cracks as possible, cut off a piece with a saw about $\frac{3}{4}$ of an inch square and perhaps $\frac{1}{4}$ of an inch thick. One of the larger surfaces is then rubbed flat on the pumice stone, keeping it well wetted with water, and then polished upon the hone, also moistened with water. Sometimes it is found to be advantageous to soak the piece of coal in a very thin solution of Canada balsam in chloroform or benzol, as directed for softer rocks (p. 70), or in a solution of shellac in spirits of wine; in either case allowing the specimen to dry thoroughly in a warm place. The polished surface is next cemented to an ordinary microscopic glass slip (3 inches by 1 inch) with the best marine glue; and this process requires care, for it is not easy to exclude all the air-bubbles, and, if they are not excluded, the section is very apt in the last stages to break away wherever they occur. The piece of coal is next reduced to about $\frac{1}{18}$ of an inch by means of a grindstone; some of the softer kinds may be cut down with a penknife. Care should be taken not to scratch the glass in the process of grinding, for most sections of coal, when once ground thin, are too fragile to allow of their being removed from the glass, but have to be covered and finished off upon the same slide. The pumice stone or corundum stick is next brought into use. The section being turned downwards, hold the glass slide between the middle finger and thumb, whilst the forefinger is placed upon the centre of the slide. In this manner the section may be rubbed round and round over every part of the pumice, using plenty of water, until it is sufficiently reduced in thickness; experience alone showing how far this process may be carried. The section is finally rubbed in a similar manner upon the hone (or water of Ayr stone). It is sometimes found necessary to use the hone even while the section is absolutely opaque, for many coals are so brittle that they crumble to pieces upon the pumice long before they show any indications of transparency. When sufficiently transparent the section may be trimmed with a penknife and the superfluous marine glue cleaned off. The section is now to be moistened with turpentine, a drop of ordinary Canada balsam (not too hard) placed upon it, and

Preparation of Sections. 73

covered in the usual way. Whatever heat is necessary should be carefully applied to the cover glass by reversing the slide for a moment or so over a spirit lamp, otherwise the marine glue may be loosened and the section spoiled. Balsam dissolved in benzol must not be used for mounting, as the benzol softens the marine glue, and a good section may in this way be destroyed.'

When no section-cutting or grinding apparatus is at hand the petrologist may sometimes gain a rough insight into the mineral composition of a rock by coarsely pulverising a small fragment, and examining the powder under the microscope. In such crude examinations levigation may occasionally be advantageous, so that the minerals of different specific gravity which compose the powder may be examined separately.

FIG. 33.

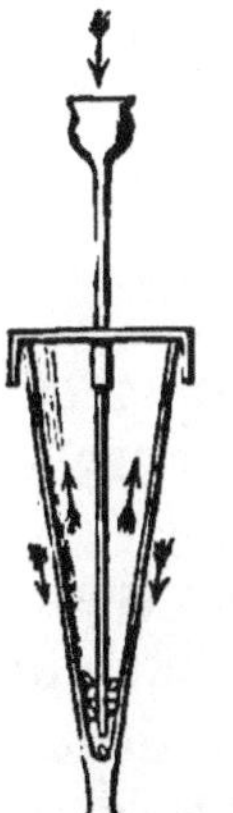

In the case of very soft rocks such as tuffs, clays, &c., useful information may sometimes be acquired by washing to pieces fragments of the rock; in this way a fine mud and often numerous minute crystals and organisms may be procured. The best apparatus for effecting this gradual washing is a conical glass about 9 or 10 inches high, across the mouth of which a cross-bar of metal or wood is fixed. A little hole drilled in the centre of the bar receives the tube of a small thistle-headed glass funnel. Roughly broken fragments of the rock should be placed in the bottom of the conical glass, and the apparatus set beneath a tap, from which a stream of water is continually allowed to run into the mouth of the funnel, the overflow trickling down the sides of the glass, which should consequently be placed in a sink. In this manner a constant current is kept up, and the fragments at the bottom of the glass are continually turned over, agitated, rubbed against one another, and gradually disintegrated. This action should be kept up, often for many days, until a considerable amount of disintegrated matter has accumulated. Samples should then be taken out by means of a pipette and examined under the microscope.

When the observer wishes to mount either such materials or fine scaly, powdery or minutely crystallised minerals, the best method is to spread a little of the substance on a glass slide, moisten the powder with a drop of turpentine, and then add a drop of Canada balsam, and cover in the usual manner. If the attempt be made to mount such substances directly in balsam, without the intervention of turpentine or some kindred medium, air-bubbles are almost certain to be included in the preparation.

CHAPTER IX.

ON THE EXAMINATION OF THE OPTICAL CHARACTERS OF THIN SECTIONS OF MINERALS UNDER THE MICROSCOPE.

It is only possible in this short chapter to give a slight sketch of some of the optical properties of minerals. The student will find the subject more fully treated in the Text-Books of Mineralogy and of Light, published in this series.

The phenomena of polarisation are also described in so many text-books of physics, some of which are specially devoted to optics and others exclusively to the study of polarisation, that for further information on the subject the reader is referred to the following works :—

'Manuel de Minéralogie,' A. Descloizeaux (Introduction), p. xxvi. t. i. 1862.

'Mémoire sur l'emploi du Microscope polarisant.' A. Descloizeaux.

'Text-Book of Mineralogy,' E. S. Dana. Wiley & Sons, New York. 1878.

'Elemente d. Petrographie,' A. von Lasaulx, pp. 13–18. Bonn. 1875.

'The Nature of Light, with a General Account of Physical Optics,' by Dr. Eugene Lommel. International Scientific Series. 1875.

'Lectures on Polarised Light,' delivered before the Pharmaceutical Society of Great Britain, by J. P(ereira). Longman. 1843.

'Mikroskopische Physiographie d. petrographisch-wichtigen Mineralien,' by H. Rosenbusch. 1873. Pp. 55–107.

'Polarisation of Light,' by Wm. Spottiswoode. Nature Series. 1874.

'Notes of a Course of Nine Lectures on Light,' by John Tyndall. 1872.

'Six Lectures on Light,' delivered in America, John Tyndall. 2nd edit. 1875.

'A Familiar Introduction to the Study of Polarised Light,' by C. Woodward. Van Voorst. 1861.

'Mikroskopische Beschaffenheit d. Mineralien u. Gesteine,' by F. Zirkel, p. 16. Leipzig. 1873.

'Physikalische Krystallographie.' Groth. Leipzig. 1876.

For the purpose of investigating the optical properties of minerals, various instruments, such as the tourmaline pincette, the dichroscope, the stauroscope, Nörremberg's polariscope, Descloizeaux's polarising microscope, Rosenbusch's stauromicroscope, &c., have from time to time been devised. The apparatus most commonly employed with microscopes consists of two Nicol's prisms, one fitted beneath the stage of the microscope and the other above the eye-piece of the instrument or above the objective, the lower one acting as the polariser, the upper one as the analyser. The analyser is sometimes so fitted as to be incapable of revolution, but the polariser is always encased in a tube in which it can freely revolve, the needful movement being imparted directly by hand, and, although rack-work is fitted to the sub-stages of some microscopes, the simpler method of turning the polariser is by far the best. It will also be found most advantageous to have the analyser fitted in a cap which slides over the eye-piece of the microscope, for it is desirable that there should be as little difficulty as possible in removing and replacing either of the Nicols.

When the principal optical sections of the two Nicols, *i.e.* when the shorter diagonals of the two prisms coincide in direction, the field of the microscope appears clear and well-illuminated. When, however, they are set at right angles to one another, there should be a total extinction of light, the field appearing perfectly dark. In intermediate positions the field becomes more or less obscure, the obscurity increasing as the principal sections of the Nicols approximate to an angle of 90°. It is advantageous to have the means of setting the two Nicols accurately at right-angles.

Supposing the Nicols to be crossed and the field to be quite dark, a slip of glass or a thin slice of rock-salt, if placed on the stage of the microscope, will produce no change, the field still remaining dark. This is owing to the glass and the rock-salt being singly-refracting or isotropic substances, the one amorphous and the other crystallising in the cubic system. All truly amorphous substances, *i.e.* those in which no crystalline structure is developed, are singly-refracting, so long as they are not subjected to conditions of abnormal strain or sudden and unequal changes of temperature resulting in corresponding molecular disturbances, but since such phenomena are only observed in thick plates of glass, &c., and are engendered by artificial means, they do not specially concern the petrologist. Again, not only rock-salt but all other minerals which crystallise in the cubic system are, with one or two doubtful exceptions, singly refractive. If we interpose between crossed Nicols a thin section of a porphyritic pitchstone, the field of the microscope will no longer appear totally dark. The matrix, or the pitchstone itself, is vitreous and behaves like glass in being singly-refracting and so affords a dark ground, but the porphyritic crystals, which are frequently felspars or micas, appear brightly illuminated, and often polarise in brilliant colours. Microliths or crystals too minute to be identified with any particular mineral species, are also of common occurrence in pitchstones, forming strings or streams, their longest axes lying in more or less definite directions. These microliths also frequently polarise. Such doubly-refracting crystals and microliths, therefore, appear brightly illuminated, while the glassy matter in which they are imbedded remains dark between crossed Nicols.

All minerals which polarise, in other words which exhibit double refraction or are anisotropic, are neither amorphous, nor do they crystallise in the cubic system. It does not, however, follow, because the section of a crystal exhibits single refraction, that it therefore belongs to the cubic system; the direction in which the section is cut must be taken into consideration, because crystals belonging to the tetragonal and the hexagonal systems are singly refractive when viewed in the direction of the principal crystallographic axis, in other words when their sections coincide with their basal planes. This, however, is the only direction in which single refraction occurs in the minerals of these doubly-refracting systems; they are consequently spoken of as optically uniaxial. The crystals of the three remaining systems, viz., the rhombic, the monoclinic, and the triclinic, are optically biaxial, *i.e.* there are two directions within them along which single refraction takes place. With regard to single refraction, then, we may sum up the foregoing statements in the following manner: that when single refraction is exhibited, *i.e.* when the object in the field of the microscope remains dark between crossed Nicols, the mineral is either:

		Amorphous . Singly refractive in all directions.

<table>
<tr><td rowspan="6">Crystallised, crystalline, or having the molecular condition of a crystal.</td><td rowspan="3">Uni-axial.</td><td>Cubic . .</td><td>,,</td><td>,,</td><td>,,</td></tr>
<tr><td>Tetragonal
Hexagonal</td><td colspan="3">,, ,, the direction of the principal crystallographic axis.[1]</td></tr>
<tr><td colspan="4"></td></tr>
<tr><td rowspan="3">Bi-axial.</td><td>Rhombic
Monoclinic
Triclinic</td><td colspan="3">Singly refractive in one or the other of the two optical axes, or singly refractive when an axis of elasticity[2] coincides with the shorter diagonal of the polariser.</td></tr>
</table>

[1] The sections parallel to the basal planes of hexagonal crystals may be distinguished by the adjacent faces affording angular measurements of 120°.

[2] The following notes extracted from De la Fosse's *Cours de Minéralogie* will to some extent suffice to indicate the difference which exists between an optical axis and an axis of elasticity:—'When a ray of light falls normally upon a face which is perpendicular to a certain direction in a doubly-refracting crystal, it becomes separated into two rays which are polarised in opposite directions and which are propagated with different velocities and generally in different directions, but there are certain special directions in which this angular separation of the two oppositely polarised rays does not take place, so that the two rays appear to coalesce and to behave as a single ray; and, if the face from which they emerge be also at right angles to this line, the two rays do not separate on their emergence, and consequently an object viewed in this direction through two parallel faces of a crystal affords a single image. If, however, emergence take place from a face situated obliquely to this direction, separation of the two rays occurs and a double image is formed. The particular directions which fulfil these conditions are the *axes of elasticity*. In some doubly-refracting crystals there are only three of these axes, in others there is an infinity of them. The axes of elasticity may therefore be defined as lines along which there is no bifurcation or angular separation of the two refracted rays, under the conditions of incidence just enunciated; and the distinctive character of an axis of elasticity consists in a single image being visible through two faces perpendicular to this axis, and in a double image being seen through faces, one of which (namely, the one turned towards the object) is perpendicular to this axis of optical elasticity, and the other oblique to it. In axes of elasticity there is always a difference in the rate of propagation of the polarised rays, which is very great compared with that which takes place along other lines among rays differently polarised. In the optical axes, on the contrary, no difference exists in the rate of propagation, which is absolutely the same for all rays which pass along them, whatever may be their plane of polarisation. The optic axes are also termed axes of double refraction, since they may be regarded as the directions in which double refraction

In doubly-refracting minerals it should be observed that chromatic effects are produced by polarisation only when the section is not taken at right angles to the optical axes. When in a doubly-refracting mineral an axis of elasticity (*i.e.* a direction along which, under certain conditions, a ray of light is polarised in two opposite directions with different velocities, but without undergoing bifurcation) corresponds in direction with the shorter diagonal of the polariser, no change from obscurity is visible when the Nicols are crossed, but on revolving the object in a horizontal direction it no longer remains dark but transmits light and polarises in colours. By turning the object in a horizontal plane or in a plane at right angles to the axis of vision, it is therefore possible to distinguish singly-refracting or isotropic from doubly-refracting or anisotropic minerals. The thickness of the section under examination may, however, in some cases, especially when the divergence between the ordinary and extraordinary ray is slight, become too limited to permit any display of double refraction.

The phenomena presented by thin sections of minerals, in plane polarised light between crossed Nicols, are as follows:—

1. During a complete revolution of the preparation all the sections of the mineral remain obscure . . ISOTROPIC.
    a. The isotropic sections show no indication of crystalline structure . . . . . . *Amorphous.*
    b. The isotropic sections exhibit polyhedral boundaries, regular lines of cleavage, or other characters indicative of crystalline structure . . . *Cubic.*
2. During a complete revolution of the preparation, some sections of the mineral remain obscure, others undergo four extinctions in directions at right angles . . ANISOTROPIC.
    a. The sections which remain obscure throughout the revolution are rectangular or octagonal in form
    Tetragonal.

becomes abolished—directions in which a bifurcation of the rays is no longer induced, and in which single images only are formed instead of the double images which are produced in directions along which a separation of the rays takes place.'

 b. The sections which remain obscure throughout the revolution are regular hexagons . . *Hexagonal.*

(In the crystals of the tetragonal and hexagonal systems, the principal crystallographic axis coincides with an axis of elasticity, and with the optic axis.)

3. During a complete revolution of the preparation all the sections undergo four extinctions in directions at right angles.

 a. The directions of extinction are parallel to each of the crystallographic axes . . . *Rhombic.*

(In this system the three axes of elasticity coincide with the three crystallographic axes.)

 b. Only some of the sections undergo extinction parallel to a crystallographic axis (the orthodiagonal)

           Monoclinic.

(In this system only one crystallographic axis (the orthodiagonal) coincides with an axis of elasticity. The two other axes of elasticity may be situated in various directions either in the plane of symmetry or in a plane at right angles to that plane.)

 c. None of the directions of extinction coincide with the crystallographic axes . . . . *Triclinic.*

(In this system there is no necessary relation of the axes of elasticity to the crystallographic axes.)

A study of the systems of rings and brushes which constitute the interference figures of crystals examined by means of a convergent pencil of polarised light, either in a Nörremberg's polariscope or in a polarising microscope, such as that of Descloizeaux, is very important for the precise determination of the crystallographic system to which minerals belong;[1] but the limits of this book preclude the possibility of describing either the apparatus or the phenomena, and the student is therefore referred to the Text-Book of Mineralogy published in this series.

The determination of the crystallographic system to which a mineral belongs, and the exact position of the

---

[1] Polariscopes for this purpose are manufactured by Fuess, Berlin; Laurent and Lutz, Paris; Steeg, Homburg; Browning and Ladd, London, and some other opticians.

planes of vibration and of the axes of elasticity, are best effected by means of the stauroscope. The simplest form of this instrument is that first devised by Von Kobell. An improvement upon this form has since been made by Brezina; while, for the stauroscopic examination of thin sections of minerals under the microscope, it is only needful to have, in addition to the usual polarising apparatus, a rather thick plate of calcspar, cut at right angles to the principal axis, inserted between the eye-piece and the eye-piece analyser, and to have crossed cobwebs, for centering the object, fixed within the focus of the eye-piece. The microscope used for this purpose should, however, possess an accurately graduated and concentrically-rotating stage,

FIG. 34.

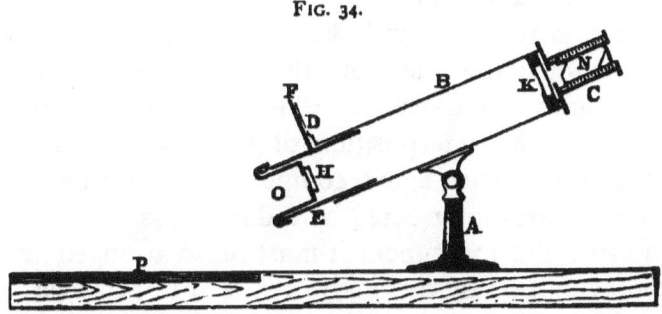

so that the angle through which the object is turned can be read off with precision.[1] It will, however, in the first place be best to give a description of the original stauroscope of Von Kobell. This consists of an upright A (fig. 34), fixed to a board, and carrying a metal tube B, which moves in a vertical plane by a joint attached to the top of the upright. The tube carries an eye-piece C, consisting of a Nicol's prism N, which serves as the analyser, and is capable of rotation about the axis of the instrument. A fixed index, D, is attached to this tube. Another tube, E, slides within the tube B

[1] Should the object fail to fill the field of the microscope, a perforated metal plate may be superposed in order to cut off the surrounding portion of the field.

G

and carries a semicircle F, which is divided into degrees. The tube E, with its graduated semicircle, revolves easily within the outer tube B, and the angle through which it is made to revolve may be read off by means of the semicircle F and the index D. Within the tube E slides another short tube, O, partially closed at its superior extremity by a diaphragm; O and E being so fitted that they turn together. Over the hole in the diaphragm of O the plate or slice of the mineral under examination is fixed with wax. A plate of calcspar, cut at right angles to the principal axis of the rhombohedron, is placed in the upper end of the tube B, just below the analyser. A parcel of blackened plates of glass, P, let into the foot-board, constitute the polariser. The light reflected from and polarised by P would, if nothing intervened between it and the Nicol N, present a perfectly dark field when the plane of vibration of the Nicol was set at right angles to the plane in which the light was polarised from P; but the interposition of the calcspar plate gives rise to an interference figure, composed of a concentric series of coloured rings intersected by a dark cross.

In using the instrument it must be so arranged that the plane of vibration of the analyser N is at right angles to the plane of vibration of the light polarised by P, and this must be effected by turning the tube E until the zero point of F lies under the index D. The object-carrier O is then taken out, and the plate of the mineral to be examined, the faces of which should be smooth and parallel, is stuck over the hole in the diaphragm with wax, but so arranged that one of its edges is placed parallel to an engraved line. Two of these lines are engraved on the upper surface of the diaphragm plate, one running from 0° to 180°, and the other from 90° to 270°. The cylinder O is then replaced in E.

Should the planes of vibration, and also the axes of elasticity in the mineral plate, be parallel to the planes of vibration of the analyser and polariser, the interference figure produced by the calcspar plate will undergo no distortion. When this is not the case the tube E, and with it the tube O and the object H, must be turned until the dark cross is perfectly restored, the amount of the revolution being read off on the semicircle F. The object-carrier is then removed and the object reversed, so that its other smooth face is turned towards the eye-piece, care, however, being taken that it is readjusted with the same edge on the same engraved line of the object-carrier. On being once more replaced, the tube E (with its contained carrier and object) is again turned until the dark cross of the interference figure is re-established, the angle of rotation being again noted, the mean of the two readings giving the inclination of the observed plane of vibration to the selected and adjusted edge of the mineral plate. Although good results are to be obtained with this instrument, it is difficult to perceive any marked change in the dark cross when the object is only turned through one or two degrees, while the removal of the object-carrier and the reversal of the object is also objectionable. To obviate these imperfections Brezina constructed a somewhat different instrument, employing a combination of two calcspar plates cut nearly at right angles to the principal axis, which affords a very sensitive interference figure, indicated in fig. 35.

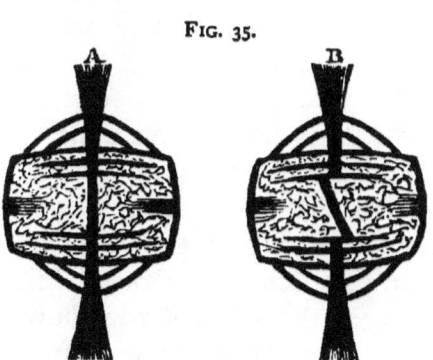

FIG. 35.

The middle band of this interference figure becomes dislocated, as shown in fig. 35 B, when the axis of elasticity in the object deviates very slightly from the principal optical section of the analyser, and, if the method of reversal of the plate be adopted, the reading of the angles of rotation, corresponding with the two displacements of the middle band, the one to the right, and the other to the left, will afford accurate results to within a few minutes.

The different crystallographic systems may be determined stauroscopically in the following way.

(1) If the plate under examination be amorphous, or if it belong to the cubic system, the cross of the interference figure remains unchanged in all positions of the plate.

(2) If the plate belong to the tetragonal or to the hexagonal system, and the adjusted edge of the plate under examination be situated parallel or at right angles to the principal axis of the analyser, the cross remains unchanged. In the case of crystals of the hexagonal system, there is also no change of the interference figure during the rotation of the object-carrier.

Should the selected edge of the plate not be adjusted in the way just described, the object must be rotated until a sharply-defined dark cross is again visible. When this is the case parallelism is established between the plane of vibration of the polariser and the principal axis, or a direction at right angles to the principal axis in the mineral under examination, since that is the position of axes of elasticity in uniaxial minerals.

(3) In the rhombic system, when the plate is arranged at zero, the cross remains unchanged in its position when any one edge of the plate parallel to a crystallographic axis is adjusted on the engraved lines of the object-carrier. If, however, the selected edge have not this position, the plate must be turned until the interference figure is restored; the angle of rotation being that which exists between that edge and an axis of elasticity.

In the rhombic system the three axes of elasticity coincide with the three crystallographic axes.

(4) In the monoclinic system the orthodiagonal alone corresponds with one of the axes of elasticity, and only in the case where an edge, parallel or at right angles to the orthodiagonal, is adjusted on the engraved guide-lines of

the object-carrier, does the interference figure remain unchanged at zero. If the adjusted edge be in any other direction, whether parallel to the vertical axis or to the clinodiagonal, the object must be turned until the cross is restored, the angle of rotation representing the angle between the adjusted edge and one of the axes of elasticity. It also gives the inclination of the crystallographic axis to the axis of optical elasticity.

(5) In the triclinic system, the axes of elasticity and the crystallographic axes bear no relation to one another; consequently, when a crystallographic axis coincides with the plane of vibration in the polariser, the interference figure appears distorted, and is only again rectified when an axis of elasticity is turned into a direction parallel to the plane of vibration of the polariser.

When an axis of elasticity in the plate of the mineral under examination lies parallel to the plane of vibration of the polariser, another axis of elasticity is at the same time parallel to the principal optical section of the analyser, since the optical sections of the two Nicols (if two Nicols be used) are at right angles. Under these circumstances the cross in the interference figure remains unchanged.

When plates of a mineral can be examined in three directions at right angles to one another, it is then easy to determine the crystallographic system of the mineral by means of the stauroscope. In microscopic sections of rocks, sections of crystals of the same mineral may often be met with in the three directions needful for a complete stauroscopic examination.

CHAPTER X.

THE PRINCIPAL ROCK-FORMING MINERALS: THEIR MEGASCOPIC AND MICROSCOPIC CHARACTERS.

THE following list comprises the minerals which usually occur as components of rocks. Many others might be added, but the limits of this work preclude the possibility of describing a larger number.[1]

(1) Species of the Felspar Group.
(2) Nepheline.
(3) Leucite.
(4) Scapolite and Meionite.
(5) Sodalite, Hauyne, and Nosean.
(6) Olivine.
(7) Hypersthene.
(8) Enstatite.
(9) Bronzite.
(10) Species of the Pyroxene Group.
(11) Species of the Amphibole Group.
(12) Species of the Mica Group.
(13) Chlorite.
(14) Talc.
(15) Tourmaline.
(16) Epidote.
(17) Sphene.
(18) Species of the Garnet Group.
(19) Topaz.
(20) Zircon.
(21) Andalusite and Kyanite.
(22) Apatite.
(23) Rutile.
(24) Cassiterite.
(25) Calcspar.
(26) Quartz, &c.
(27) Magnetite.
(28) Titaniferous iron.
(29) Hematite.
(30) Limonite.
(31) Iron and Copper Pyrites.
(32) Zeolites.
(33) Viridite, Opacite, &c.

SPECIES OF THE FELSPAR GROUP.

Felspars are essentially silicates containing alumina, together with potash, soda, baryta,[2] lime, or any two or three of these bases, which often to a certain extent replace one another in the different species. With few exceptions,

---

[1] Some short but admirable notes on the determination of the optical characters of minerals occur in a paper in the *Bulletins de la Société belge de Microscopie*, tome iv. 1877-78, entitled 'Note sur un Microscope destiné aux Recherches minéralogiques' by A. Renard, S.J.

[2] Baryta occurs in the species hyalophane, which is, however, a mineral of comparative rarity.

the felspars are white or of pale colour. Except when in a decomposing or decomposed condition, they have a hardness of about 6, *i.e.* they can be scratched, but not easily, with the point of a knife.

Chemically they may be divided into three groups: the alkali, the lime, and the mixed alkali-lime or alkali-baryta felspars.

Crystallographically they are represented in two systems: the potash and potash-baryta felspars occurring in the monoclinic, and the others in the triclinic or anorthic system.

Before the blowpipe the potash felspars fuse with difficulty, while the soda felspars fuse more readily. Both are insoluble in acids, except hydrofluoric acid. The principal lime felspars, labradorite and anorthite, are both of them soluble in acids. Labradorite fuses readily before the blowpipe, while anorthite is more difficultly fusible: the former fusing at 3, the latter at 5, of Von Kobell's scale.

The two principal directions of cleavage in the monoclinic felspars are at right angles to one another; those of the triclinic felspars lie at angles other than 90°.

Both the monoclinic and triclinic felspars have a perfect cleavage parallel to the basal plane. The other perfect cleavage is, in the monoclinic felspars, parallel to the clinodiagonal; in the triclinic it is parallel to the brachy-diagonal. In both systems there are hemiprismatic cleavages which are more or less imperfect, and more perfect in one direction than in the other.

When the light falls somewhat obliquely on the basal cleavage plane of a triclinic felspar it is usually seen to be traversed by numerous parallel striations, the interspaces between the striæ representing twin lamellæ. This twinning is frequently so many times repeated in the felspars of this system that more than fifty lamellæ have been noted, under the microscope, in a single crystal.[1]

[1] 'Mikroskopische Untersuchungen über Diabase.' *Zeitsch. d deutsch. geol. Gesclsch.*, Bd. xxvi. Heft i. p. 1, by J. F. E. Dathe.

The accompanying diagrams (fig. 36) represent how these

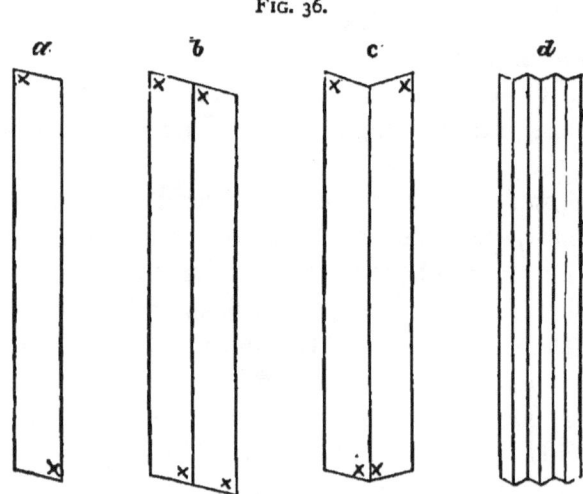

striæ are due to hemitropy, while fig. 37 shows the appearance of the twin lamellæ unler polarised light.

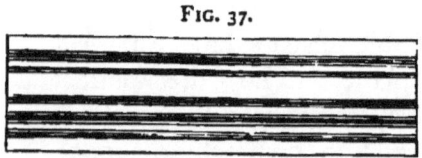

The absence of this striation, however, must not always be taken as an indisputable proof that the felspar is not triclinic: nevertheless it is most common in, and characteristic of, the felspars of this system.

When the light falls obliquely either on the basal plane, the orthopinakoid, or the hemidome of a monoclinic felspar, a simple twinning, as evinced by difference of lustre, may often be noticed in the crystals; this twinning takes place upon the type known as the 'Carlsbad type.' Twinning upon other, but far less frequently occurring, types is, however, also known, and will be described in the following pages.

In the opinion of Tschermak, the species albite and anorthite are isomorphous, the soda in albite being represented by lime in anorthite: intermediate variations occurring between the two species in the percentage of silica; six molecules of silica occurring in the formula of albite, and only two in that of anorthite.

The following tables, showing the formulæ, the percentage composition, and the oxygen ratios of the different species, will assist the student in learning the relations which they bear to one another.

CHEMICAL FORMULÆ OF THE PRINCIPAL FELSPARS.

Orthoclase $K_2O \cdot 3SiO_2 + Al_2O_3 \cdot 3SiO_2$

or $\dfrac{K_2O}{Al_2O_3}$ $6SiO_2$

Albite $\quad Na_2O \cdot 3SiO_2 + Al_2O_3 \cdot 3SiO_2$

or $\dfrac{Na_2O}{Al_2O}$ $6SiO^2$

Anorthite $CaO \cdot SiO_2 + Al_2O_3 \cdot SiO_2$

or $\dfrac{CaO}{Al_2O_3}$ $2SiO_2$

According to Tschermak's view 3 molecules of albite + 1 molecule of anorthite constitute oligoclase:

$$\begin{array}{r} 3Na_2O + 3Al_2O_3 + 18SiO_2 = 3 \text{ of albite} \\ \underline{CaO \ldots \ldots + Al_2O_3 + 2SiO_2 = 1 \text{ of anorthite}} \\ CaO + 3Na_2O + 4Al_2O_3 + 20SiO_2 = \text{oligoclase} \end{array}$$

1 molecule of albite and 1 molecule of anorthite constitute andesine:

$$\begin{array}{r} Na_2O + Al_2O_3 + 6SiO_2 = 1 \text{ of albite} \\ \underline{CaO \ldots \ldots + Al_2O_3 + 2SiO_2 = 1 \text{ of anorthite}} \\ CaO + Na_2O + 2Al_2O_3 + 8SiO_2 = \text{andesine} \end{array}$$

1 molecule of albite + 3 molecules of anorthite constitute labradorite:

$$\begin{array}{r} Na_2O + Al_2O_3 + 6SiO_2 = 1 \text{ of albite} \\ \underline{3CaO \ldots \ldots + 3Al_2O_3 + 6SiO_2 = 3 \text{ of anorthite}} \\ 3CaO + Na^2O + 4Al_2O_3 + 12SiO_2 = \text{labradorite} \end{array}$$

PER-CENTAGE COMPOSITION OF THE PRINCIPAL FELSPARS.

Orthoclase $SiO_2 = 64·20$ $Al_2O_3 = 18·40$ $K_2O = 16·95$
Albite $SiO_2 = 68·6$ $Al_2O_3 = 19·6$ $Na_2O = 11·8$
Anorthite $SiO_2 = 43·1$ $Al_2O_3 = 36·9$ $CaO = 20$
Oligoclase $SiO_2 = 62·1$ $Al_2O_3 = 23·7$ $Na_2O = 14·2$
Labradorite $SiO_2 = 52·9$ $Al_2O_3 = 30·3$ $CaO = 12·3$ $Na_2O = 4·5$
Andesine $SiO_2 = 59·7$ $Al_2O_3 = 25·6$ $CaO = 7$ $Na_2O = 7·7$

OXYGEN RATIOS OF THE PRINCIPAL FELSPARS.

Orthoclase. Albite. Oligoclase. Andesine. Labradorite. Anorthite.

  1 : 3 : 12   1 : 3 : 12   1 : 3 : 10   1 : 3 : 8   1 : 3 : 6   1 : 3 : 4

being respectively the oxygen ratios for $RO . R_2O_3$ and SiO_2.

$RO = CaO$ Na_2O and K_2O. and $R_2O_3 = Al_2O_3$.

In this manner Tschermak limits the number of species by regarding labradorite, oligoclase, and andesine as admixtures in different proportions of the two species albite and anorthite.

The following extract from Dana's 'System of Mineralogy,' 5th ed., 1871, p. 336, may here be cited, as pointing out the intercrystallisation which probably gives rise to the compound-specific character of some felspars. After giving a list of the oxygen ratios for the different species, he adds: 'The species appear in the analyses to shade into one another by gradual transitions; but whether this is the actual fact, or whether the seeming transitions (when not from bad analyses) are due to mixtures of different kinds through contemporaneous crystallisation, is not positively ascertained. The latter is the most reasonable view. It has been shown by Breithaupt and others that orthoclase and albite (or the potash and soda feldspars) occur together in infinitesimal interlaminations of the two species, and that the soda-potash variety, perthite, is one of those thus constituted. This structure is apparent under a magnifying power, and also when specimens are examined by means of polarised light. Moreover, these and other feldspars very commonly occur side by side or intercrystallised when not interlaminated, as oligoclase and orthoclase in the granite of Orange Summit, N. Hampshire, and Danbury, Conn.; in obsidian in Mexico; in trachytes of other regions. Such facts show that the idea of indefinite shadings between the species is probably a false one, since the two keep themselves distinct, and, in the perthite and similar cases, even to microscopic perfection. They also make manifest that contemporaneous crystallisation is a true cause in many cases.'

The following is a short description of the characters of the different species of felspar commonly occurring as constituents of rocks. They are here divided into two groups according to the systems in which they crystallise; since it is at present a matter of considerable difficulty to discriminate between the different species of the triclinic system, especially when the crystals are so minute as to be incapable of isolation for the purposes of chemical analysis, and since all the felspars which crystallise in the triclinic system present approximately the same microscopic appearances under polarised light. The two groups into which they are classed for the present purposes of the petrologist, especially when regarded microscopically, are the orthoclastic ($\mathrm{\dot{o}} \rho \theta \mathrm{\acute{o}} \varsigma$ and $\kappa \lambda \mathrm{\acute{a}} \omega$ = rectangular cleavage), or that in which the chief cleavages are mutually situated at right angles, and the plagioclastic ($\pi \lambda \mathrm{\acute{a}} \gamma \iota o \varsigma$ and $\kappa \lambda \mathrm{\acute{a}} \omega$ = oblique cleavage), or those in which the cleavage planes intersect at angles other than 90°. These two groups, whose members are respectively spoken of as orthoclase and plagioclase, may in most cases be readily distinguished under polarised light by the differences which they present in their twinning; crystals of orthoclase and its varieties usually showing, when twinned, a median divisional plane, on either side of which the halves of the crystals depolarise the light in complementary colours; while in the case of plagioclase the crystals exhibit numerous bands of different colours. When sections of plagioclase are ground very thin, their twin lamellæ usually present only pale blue or neutral tints under polarised light; when thicker, strong colours, often variegated, mark the different lamellæ. The student is, however, here warned that conditions may occur, or that sections of crystals may be so cut, that these phenomena, although they may exist, are not rendered apparent. Doubtful cases also occur in which, at times, it is very difficult to assign a felspar crystal with absolute certainty either to the one system or the other.

ORTHOCLASE.

Crystalline system monoclinic or oblique. The following figures (38 and 40) represent the common forms. Fig. 39 shows a crystal twinned on the Carlsbad type. Figs. 42, 43, and 44 represent the twinning of fig. 41 upon the Baveno type; the lines T T indicating the planes of composition. The formulæ on the different faces are those of Naumann. In the variety sanidine the orthopinakoid is usually less developed than in common orthoclase.

Angle of oblique rhombic prism 118° 48'. Cleavage parallel to the base and clinodiagonal very perfect and at right angles; parallel to one or other hemiprism imperfect.

In polarised light under the microscope the crystals sometimes exhibit moderately strong colours. Crystals twinned on the Carlsbad type are of common occurrence; and, when the plane of section coincides with the orthopinakoid, they polarise in different colours on either side of a median line which represents the plane of composition, the difference in colour being due to the difference of direction of the optical axes in the opposite halves of the crystal and the positions of the planes of chief vibration in the Nicol's prisms of the polariscope, the difference of direction of these axes being due to hemitropy or a half revolution of one of the halves of the crystal. When the section is cut more or less obliquely to the orthopinakoid the divisional line approaches nearer to one side or the

Orthoclase. 93

other of the crystal. When the sectional plane almost coincides with the clinopinakoid only a narrow marginal

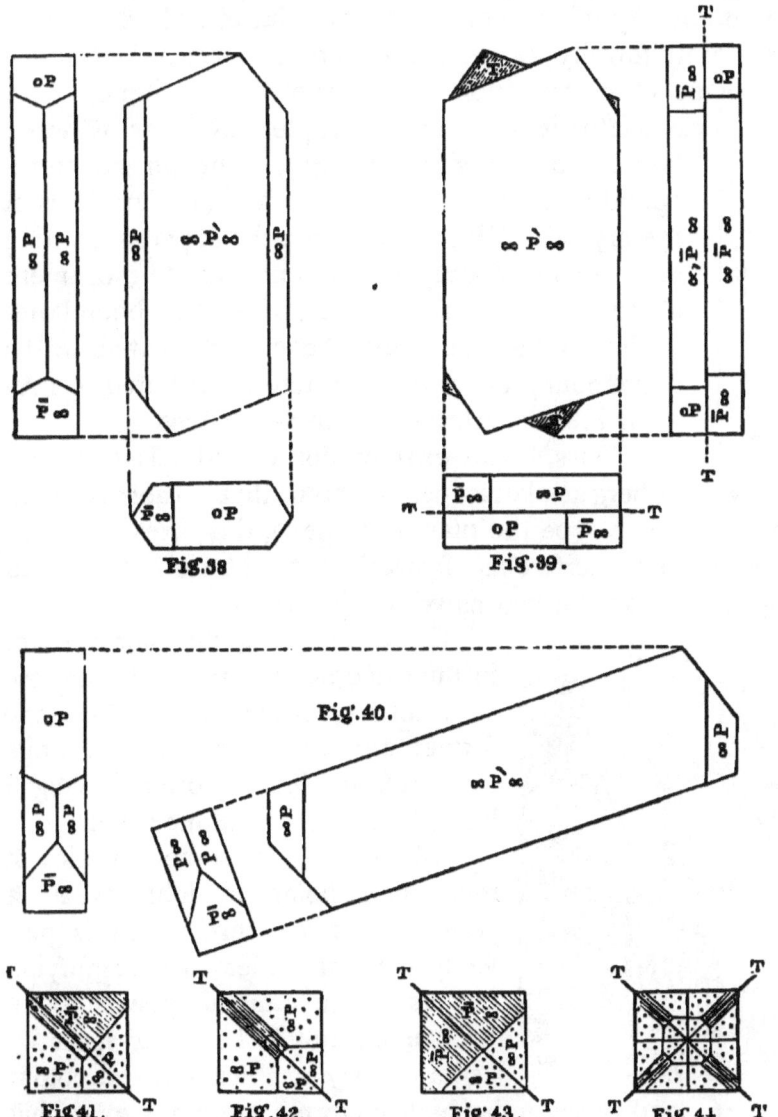

Fig. 38   Fig. 39.

Fig. 40.

Fig. 41.   Fig. 42.   Fig. 43.   Fig. 44.

band of a different colour represents one of the halves of the crystal, and, when it coincides completely with the clinopinakoid all signs of twinning are suppressed, and the crystal presents a uniform sheet of colour. Variations in the uniformity of this colour are then due to a corresponding want of uniformity in the thickness of the section. Usually the plane of composition is represented by a straight line, but occasionally it appears interrupted and, as it were, faulted to one side (fig. 45), although no corresponding break is visible in the boundary lines of the crystal. This implies that the apparent shifting of the plane of composition is not due to movement along a line of fracture; for, if so, the boundaries of the crystal would also have participated in the movement; but it must rather be attributed to irregular interpenetration of the two halves.

FIG. 45.

Weiss, by an examination of orthoclase crystals in a Nörremberg's polariscope, has shown that in those twinned on the Baveno type the planes of the optical axes (as indicated by the interference figures) stand at right angles to one another in the two halves of the crystals.

FIG. 46.

Microcline, Arendal × 50. (Polar.)

Crystals of sanidine mainly differ in microscopic appearance from those of ordinary orthoclase in that the former are clear and pellucid, while the latter are less so, often having a hazy, turbid, or nebulous aspect.

Microcline.—In some varieties of microcline, polarised light reveals a very peculiar structure which is perfectly evident under low magnifying powers. The mineral appears to be broken up into a chequered mass by septa of varying thickness and colour (fig. 46), and these under higher powers present a somewhat pectinate appearance. The septa or striæ run at right angles to one another. It is not certain that similar structure does not occur in orthoclase: thus Stelzner has pointed out that one of these directions of striation is parallel to the ortho-, the other to the clino-pinakoid. If this be the case, they do not bear any direct relation to the cleavage of the mineral, as might at first sight be imagined, and an examination of the broken edges of thin sections does not seem to lend much support to such a supposition. This structure, although pre-eminently characteristic of massive microcline, is not so frequently seen in the small crystals imbedded in rocks.[1] It is, however, often to be observed in little microscopic patches in some of the triclinic felspars, notably in oligoclase (fig. 47), thereby indicating that these minerals, as suggested by Sterry Hunt[2] and Tschermak,[3] are by no means homogeneous. Actual interlamination of albite and microcline can be seen with the naked eye in the variety of felspar known as perthite, but in most of the other felspars in which these admixtures of species occur the microcline does not, as a rule, form laminæ, but merely lies in disconnected, irregular, or lenticular patches which, however, are often disposed more or less in the general direction of twinning which the species, serving as matrix, presents. In minute sanidine crystals occurring in some vitreous and trachytic

FIG. 47.

Oligoclase, Twedestrand, Norway ×115. (Polar.)

[1] Some of the felspar crystals in the hornblendic granite, of which Cleopatra's needle is made, show this structure very distinctly.

[2] *Chemical and Geological Essays*, 2nd edition. Salem, 1878, p. 443. Sterry Hunt's conclusions, which are almost identical with those of Tschermak, were first published in the *American Journal of Science*, in Sept. 1854, or ten years previous to the appearance of Tschermak's paper.

[3] *Sitzungsberichte d. Kais. Akad. d. Wissenschaft.* Wien. Bd. I. Abth. i. 571 (1864).

rocks other peculiar structures are sometimes exhibited. These consist in markings which are usually very faint, and often necessitate the use of tolerably high magnifying powers in order to make them out clearly. In such crystals divisional markings or internal boundaries may be discerned (fig. 48). These consist sometimes of two curved lines, with their convex aspects directed inwards, and often approximating or being in actual contact, the lines seeming to spring from the lateral edges of the crystal, while at others they consist of a straight median line or rib which traverses the crystal for some distance, and then suddenly bifurcates, the bifurcations passing in straight lines to the opposite lateral edges or corners of the crystal section. These larger divisional markings, whether curved or rectilinear, are crossed by other markings or striæ, which are usually very numerous, often extremely delicate, and always observe definite directions; but, as the angles made by their intersections vary very considerably in different crystals, it is unsafe at present to hazard any conjectures as to the relation which they bear to general recognised crystalline structure.[1]

FIG. 48.

Sanidine crystal in rhyolite. Berkum, Rhine, × 115. (Polar.)

In the blue, chatoyant microcline felspar of the zircon syenite of Norway the angles of intersection of the principal cleavages are, according to Breithaupt, 90° 22′ to 90° 23′, so that in this respect it differs slightly from the cleavage of orthoclase.

Descloizeaux has found this mineral to be optically similar

---

[1] 'Notes on some Peculiarities in the Microscopic Structure of Felspars.' F. R. *Quarterly Journal Geological Society*, p. 479, 1876.

to orthoclase and he has appropriated Breithaupt's name to the green felspar know as Amazon-stone, which in certain varieties, especially those from Colorado and Arkansas, exhibits optical properties incompatible with monoclinic symmetry, while in other physical respects, and chemically, it is not distinguishable from orthoclase. These facts are of special interest as more completely establishing the isomorphism of orthoclase and albite, a pure potash felspar of triclinic symmetry having been previously unknown.

Descloizeaux has pointed out[1] that thin sections of Amazon-stone, when magnified, are seen to inclose bands and patches of albite, yet although such albite inclosures are very numerous and comparatively large in some examples, the percentage of soda which they yield on analysis is always slight.

Perthite, which presents a well-marked interlaminated structure to the naked eye, consists of differently coloured alternating bands of orthoclase or microcline, and albite.

ALBITE.

Crystalline system triclinic or anorthic. The cleavages, which are parallel to the base and brachypinakoid, intersect at angles of 86° 24' and 93° 36'. The cleavage faces usually have a pearly lustre.

A fine lamellar or twinning striation is often visible on the basal plane. The plane of composition is parallel to the brachypinakoid. Untwinned crystals are rare. In chemical composition it is essentially a soda-felspar. It is usually white or greyish.

Albite, on losing its alkaline constituents, passes into kaolin, &c., just as orthoclase does. Before the blowpipe it fuses with difficulty to a whitish glass, and colours the flame yellow. It is not acted upon by acids.

Sections of crystals, so long as they do not coincide with,

---

[1] *Annales de Chimie et de Physique*, 5me Série, t. ix. 433, 1876.

or approximate very closely to, the plane of the brachypinakoid, exhibit under the microscope coloured parallel bands when viewed by polarised light. Sections of albite seldom contain any microliths or other inclosures.

Figures 49 and 50 illustrate the albite type of twinning, while in fig. 51 the direction of the plane of composition in the twinning of pericline is shown. The lines TT indicate

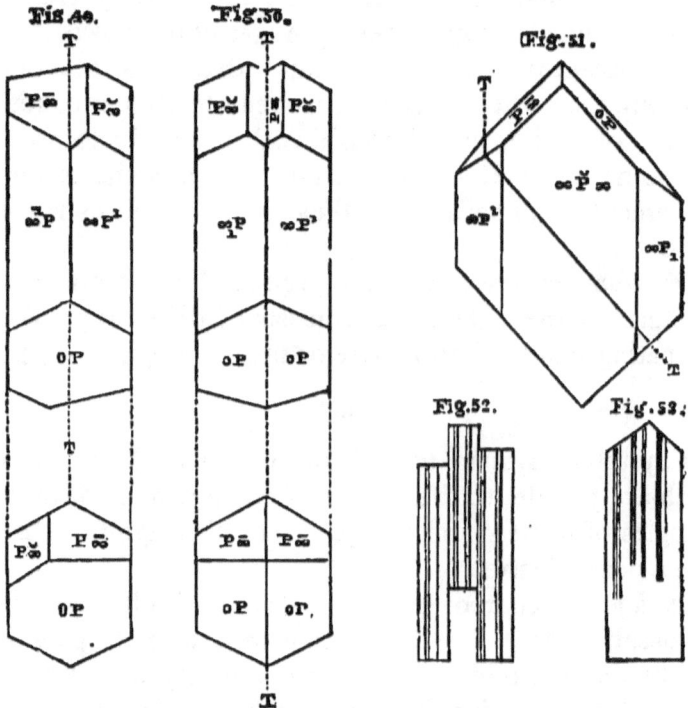

the planes along which the twinning takes place. Fig. 52 shows a group of three twinned crystals of triclinic felspar, such as are often seen under the microscope. Fig. 53 shows the abrupt termination of twin lamellæ occasionally to be observed in plagioclastic felspars.

ANORTHITE.

This is a mineral which is not of very frequent occurrence.

The best crystals are found in the Vesuvian lavas.

Its crystalline system is triclinic or anorthic.

It cleaves parallel to the basal plane and to the brachypinakoid; the cleavages intersecting at an angle of 94° 12′, and exhibiting a pearly lustre. Twin striation is not as a rule strongly marked on the basal cleavages. The plane of twinning corresponds with that of albite.

Before the blowpipe it fuses with difficulty to a clear glass. It is completely soluble in concentrated hydrochloric acid. When imbedded in rocks the crystals usually have a greasy lustre, but when formed in druses they are generally glassy and limpid.

OLIGOCLASE, ANDESINE, AND LABRADORITE.

All of these species crystallise in the triclinic system, and are traversed by two sets of cleavage planes, the one parallel to the basal plane, the other parallel to the brachypinakoid, which intersect at angles of 86° 10′ and 93° 50′. In colour they vary from white to different shades of grey, while labradorite is sometimes almost black, as that from Hamilton Sound, Labrador. Labradorite also shows a fine play of variegated colours on planes parallel to the brachypinakoid, and occasionally oligoclase likewise exhibits a play of colour.

The lustre of oligoclase and andesine is vitreous, approaching to greasy on the basal cleavage, while that of labradorite is vitreous and somewhat pearly.

Oligoclase is not acted upon by acids, except hydrofluoric acid. It fuses more readily than orthoclase before the blowpipe, and colours the flame yellow.

Andesine is only imperfectly acted upon by acids, except hydrofluoric acid. The edges of thin splinters may be fused before the blowpipe.

Labradorite in a fresh, unaltered condition is only imperfectly soluble in hydrochloric acid, but weathered samples are completely decomposed by it, with separation of gelatinous silica.

The three species of felspar now under consideration present such closely analogous, or, according to observations hitherto made, identical, characters under the microscope, that by this means alone it is at present impossible to discriminate between them, and consequently they are all described under the general term 'plagioclase' by microscopists.

By one or two ready methods of analysis, which have been devised for the purpose of identifying very minute quantities, and of superseding the more tedious processes of ordinary chemical analysis, the different species may, however, be identified. Amongst these methods may be cited that of Szabo, based upon the relative duration of colour imparted to the flame of a Bunsen's gas jet by two assays of known weight, and which may be defined as a system of colour-comparison; and a new method of chemico-microscopic investigation recently introduced by Prof. Bořicky, of Prag, which consists in treating the substance with hydro-fluo-silicic acid and examining under the microscope the different crystalline forms of the fluo-silicates produced by the use of this reagent;[1] a system resting upon the determination of the forms and chemical composition of artificially produced crystals. In cases where these artificially formed crystals belong to the same crystallographic system, but may differ in chemical composition, reagents other than hydro-fluo-silicic are employed, and the crystals formed by these subsequent reactions give confirmatory evidence of the chemical nature of the previously undetermined fluo-silicates, which resulted from the first reaction with hydro-fluo-silicic acid.

Prof. Bořicky informs the author that he has recently been conducting experiments upon this system on minute

---

[1] 'Elemente einer neuen chemisch-mikroskopischen Mineral- und Gesteinsanalyse. Bořicky (*Archiv d. Naturw. Landesdurchforschung von Böhmen. III. Band. Chem.-petrologische Abtheilung*), 1877.

sections of minerals occurring in thin microscopic slices of finely crystalline rocks, and that from fragments or imbedded crystals not more than 0·2 millemeters to 0·7 millemeters square, he has procured good results, which 'are often remarkably beautiful, so that even minerals in very fine-grained rocks may be separately examined.'[1]

The limits of this work preclude the possibility of anything more than a brief allusion to this method of investigation.

The hydro-fluo-silicic acid used for this purpose is made in a leaden retort from fluoride of barium, sulphuric acid, and pure powdered quartz; the resulting fluo-silicate is then transferred to a platinum dish containing water, and after moderate dilution is decanted into a gutta-percha bottle. Minute quantities of the reagent are applied to the minerals under examination by means of a gutta-percha rod terminated by a little spoon-shaped groove. The strength of the solution used by Prof. Bořicky is about $3\frac{1}{2}$ per cent. It is important that neither too weak nor too strong a solution be used, since in the former case many minerals do not afford any satisfactory reactions, and in the latter so many crystals of fluo-silicates are formed, and from many silicates so much silica is separated, that the field of the microscope becomes filled with an indistinct, hazy mass, in which no definite crystalline forms can be distinguished. In such a case further dilution with one or two drops of water is necessary.

The fragment under examination may be about the size of a pin's head or a millet seed. A drop or two of Canada balsam should be placed on a glass slip and heated, but not to ebullition, and the slip should then be turned about so as to run the balsam into a thin even sheet. Upon this surface, when cooled and hard, is placed the small fragment of the mineral to be examined, and the slide is again sufficiently heated to cause adhesion of the fragment. A drop of the hydro-fluo-silicic acid solution should then be applied, and the slide set aside in a horizontal position

[1] Private communication from Prof. Bořicky.

on a flat plate and in a place free from dust, a small capsule containing sulphuric acid being placed beside it, and the whole covered with a small glass shade or an inverted tumbler; a dry atmosphere is thus insured, but in spite of this it often takes 24 hours before the drop is completely evaporated. When evaporation is over, the preparation is ready for examination under the microscope.

Crystals of fluo-silicate of sodium are cubic; those of magnesium, iron, and manganese are hexagonal or rhombohedral; those of lithium, strontium, and calcium are monoclinic.

The fluo-silicates of sodium, magnesium, and calcium are so different in form that they can at once be distinguished from one another. Again, the fluo-silicates of lithium differ sufficiently from those of strontium and calcium to render their recognition easy. The distinction between the crystals of fluo-silicates of calcium and strontium, and between those of magnesium, iron, and manganese is, however, scarcely possible, and other reactions must be had recourse to in order to determine their respective chemical natures. Thus, for example, if the fluo-silicates of calcium and strontium be treated with somewhat dilute sulphuric acid, the crystals of the former substance will, after a few seconds, become surrounded with a thick fringe of monoclinic, acicular crystals of gypsum, while the crystals of fluo-silicate of strontium pass slowly into granular masses, interspersed with short needles (of celestine?), a process which usually takes several hours.

In polarised light, under the microscope, crystals of plagioclase show, as already stated, a series of parallel bands or twin lamellæ, which polarise in various colours; and this appearance of plagioclase is very characteristic so long as the crystals are not cut parallel to the brachypinakoid.

Sections of labradorite crystals cut parallel to the macropinakoid sometimes show another system of striations or interlamellar growths; these form outcrop striæ on the brachypinakoid, while the ordinary twin lamellæ crop out on, and striate, the basal plane. The latter represent twinning upon the albite type, and the others possibly indicate the pericline type of twinning. Sections taken parallel to the basal plane would show the former but not the latter; sections parallel to the brachypinakoid would show the latter but not the former; while in sections parallel to the macropinakoid both systems of lamellæ would be visible, and their true angle of intersection, which, according to Dr. A. von Lasaulx, is 86° 40′, could be correctly measured.[1] These lamellæ are usually sufficiently broad, and those pertaining to the pericline type of twinning are generally so irregularly developed, or their planes of demarcation are distributed at such wide intervals, that there is little fear of mistaking such intersecting twinning-planes in labradorite for the rectangular cross-hatching which occurs in orthoclase. If the labradorite crystals were very small, and their sections did not offer the means of measuring precisely the angle of intersection of the two sets of lamellæ, some such doubt might be entertained, but it so happens that this twinning upon the pericline type is as a rule only developed in plagioclase crystals, which, microscopically speaking, are of considerable dimensions; while on the other hand the cross-hatching in orthoclase is developed even in very microscopically minute crystals and interstitial patches.

In sections of oligoclase minute imbedded patches of orthoclase may frequently be recognised under the microscope, the orthoclase exhibiting the characteristic cross-hatching in polarised light. These patches are usually very irregular in form, but are often distributed in more or less rudely parallel lines, corresponding with the direction of the twin lamellæ in the plagioclastic portion of the mineral.[2]

A method of determining the different species of triclinic felspars by the position of their axes of elasticity, is described by Max Schuster in vol. lxxx. Sitzb. d. k. Akad.,

---

[1] *Elemente der Petrographie.* A. von Lasaulx. 1875, p. 44.
[2] Tschermak, *Sitzungsberichte d. Kais. Akad. d. Wiss.* Bd. I. Abth. i. 571. Wien, 1864. Also *Q. J. G. S.*, vol. xxxi. p. 479. F. R.

Pt. 1, July 1879. The different species are characterised by the inclination of the direction of extinction to the edge formed by the faces oP and $\infty \breve{P} \infty$. On plates cleaved parallel to oP, they are seen to range from $+4°$ in albite to $-38°$ in anorthite, while on $\infty \breve{P} \infty$ the divergence is from $+18°$ in albite to $-40°$ in anorthite. (For tables, see Appendix, p. 309.)

NEPHELINE.

This mineral, which is of common occurrence in many lavas, and notably represents the felspathic constituents of many basalts, is one of the essential components of phonolite. It is very closely allied to the felspars in its chemical composition $Al_2O_3\ SiO_2 + RO\ SiO_2$. RO representing soda and potash, the former averaging about 16 or 17 per cent., and the latter about 5 per cent.

Nepheline has a conchoidal or uneven fracture and a hardness of 5·5 to 6.

Before the blowpipe it fuses to a colourless glass, and when powdered and treated with acids it gelatinises.

Nepheline crystallises in the hexagonal system in rather stout prisms which are terminated by basal planes, and are frequently modified by planes of the hexagonal pyramid. There are two directions of cleavage, one basal, the other prismatic, but neither of these cleavages is perfect. The crystals are generally colourless, or with a slight tinge of green, yellow, or brown, and have a vitreous or greasy lustre.

FIGS. 54 & 55.

Sections of nepheline crystals when cut parallel to the vertical axis afford rectangular forms whose boundaries are constituted by the planes ∞P and oP, while those cut at right angles to the vertical axis, *i.e.* parallel to the basal plane, appear as well-defined hexagons. Sections cut obliquely

Sections of Hexagonal Prism.

Plate I.

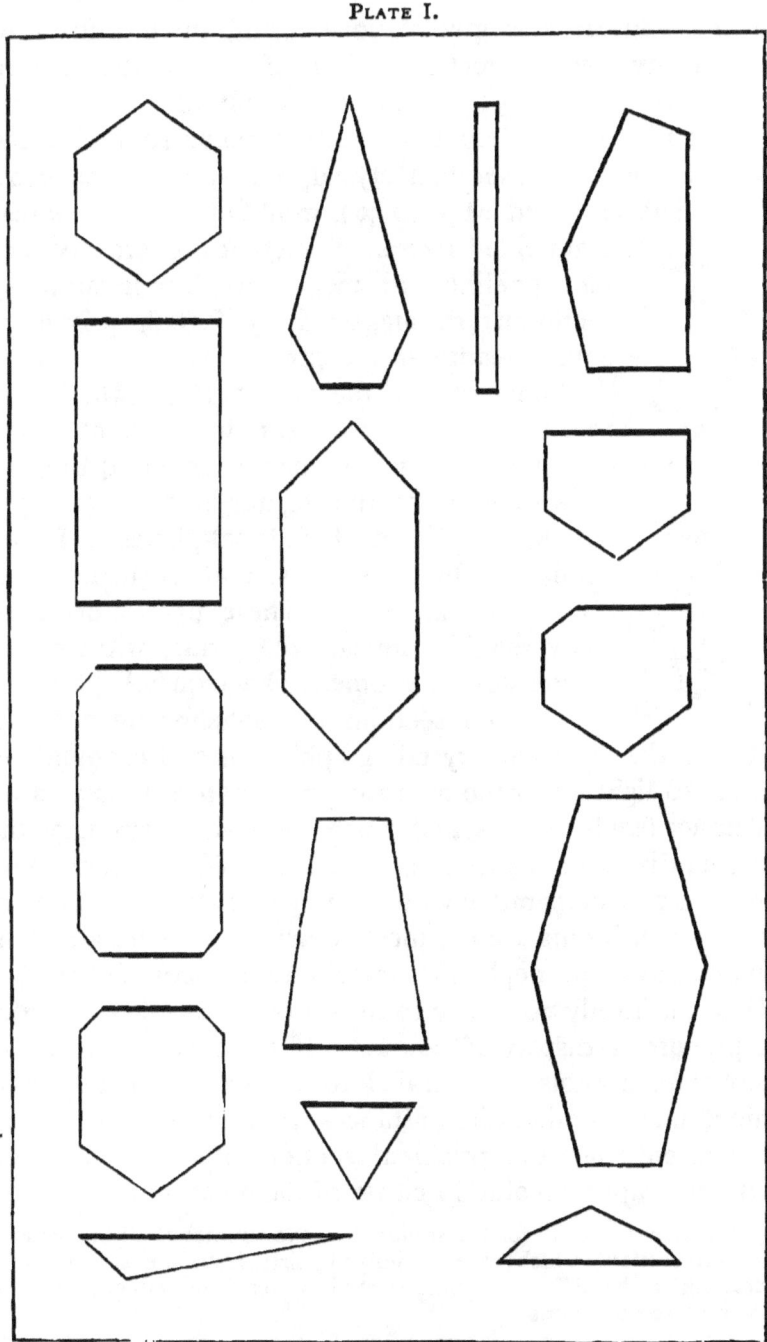

to the vertical axis give various forms, often difficult to refer to any precise direction. Thus, if a section be taken in the plane S S (figs. 54 and 55), the resulting figure will be bounded by eight sides; but a section taken parallel to the vertical axis of a modified crystal consisting of the combination P, ∞ P and oP (fig. 56), would also give a similar eight-sided figure. Sometimes square sections of nepheline are met with. These probably represent, as suggested by Zirkel, prisms of equal breadth and length.

FIG. 56.

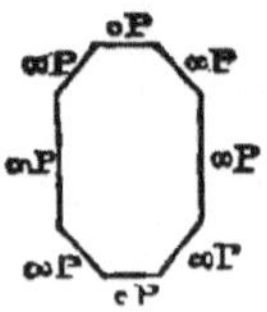

Some idea of the various forms which may result from sections taken in different directions through a hexagonal prism may be acquired from an examination of the figures in the accompanying Plate I., in which basal planes (o P) are indicated by thick lines, and prismatic faces (∞ P) by thin lines. These figures, however, only roughly indicate the forms, without any pretension to geometrical accuracy.[1]

FIG. 57.

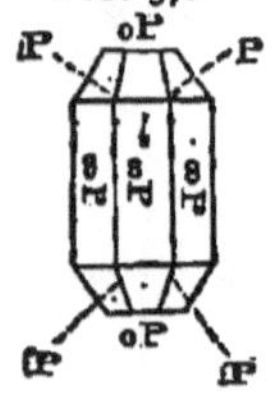

If thin sections of nepheline be cut parallel to the principal crystallographic axis and examined in polarised light under the microscope, they are seen to polarise in rather feeble colours, and which are never so strong as the colours displayed by quartz, since the double refraction of nepheline is comparatively weak; greyish, pale blue, yellowish, and brownish tints are the most common. Should a rectangular section of nepheline remain dark between crossed Nicols, it is only necessary to turn it on its own axis in order to procure a display of colour. If transverse sections of nepheline crystals cut parallel to the basal planes be examined under similar circumstances they will merely appear translucent when the principal sections of the Nicols coincide, and, upon revolution either of the polariser or analyser,

---

[1] It may be useful for the student to make outlines of all the possible forms of sections which may be derived; first, from the simple forms occurring in the different crystallographic systems and, afterwards, from observed combinations.

they will gradually darken until the principal sections of the Nicols are crossed when, if the sections be truly parallel to the basal planes, perfect obscurity sets in. Fluid lacunæ and other inclosures are of frequent occurrence in nepheline.

Among the latter pale greenish or yellowish microliths and granules of augite are common. Microliths of hornblende are comparatively rare, although they do occur in the nephelines of some phonolites.

These microliths are usually seen to lie in more or less distinct zones, corresponding with the boundaries of the sections in which they occur, but in the case of some of the larger, transverse, hexagonal sections they have been observed to assume three directions corresponding with the lateral axes of the crystals.[1] A fine dusty matter of a brown or bluish colour is often met with in nepheline, especially when it occurs in hornblendic rocks. Under a high magnifying power the dust is seen to consist of very diminutive microliths, air bubbles, and glass lacunæ. This dust is often very densely accumulated in zones which run parallel to the hexagonal boundaries of transverse sections, while, at times it is also segregated in the centre of the crystals. Nepheline crystals with dark centres, which often occupy almost the entire area of a transverse section, may be seen in the phonolite of the Wolf Rock, which lies off the coast of Cornwall.[2] The fluid inclosures in nepheline have in some cases been regarded as liquid carbonic acid, and at times they contain movable bubbles and also minute crystals; those observed by Sorby in nepheline from Vesuvius were cubic, and were considered to be either chloride of sodium or of potassium.[3] The alteration visible in weathered crystals of nepheline consists of a fibrous pale yellowish substance, which is first developed in the exterior portions of the crystals,

---

[1] Zirkel, *Mik. Beschaff. d. Min. und Gest.*, p. 143.
[2] This rock was first described by S. Allport in the *Geological Magazine*, Decade I. vol. viii. p. 247.
[3] Sorby, 'On the Microscopical Structure of Crystals,' *Q. J. G. S.*, vol. xiv. p. 480.

and gradually extends inwards. The ultimate result of such alteration is a zeolitic substance which is possibly natrolite.[1]

Elæolite is a greenish, brownish, sometimes reddish variety of nepheline. It has an oily lustre and fuses more readily before the blowpipe than nepheline. It seldom or never occurs crystallised as a rock constituent but in a massive or simply crystalline condition. As observed by Rosenbusch, it bears the same relation to nepheline that orthoclase does to sanidine. Under the microscope it is seen to contain numerous greenish inclosures, some pulverulent and referred to diaspore by Scheerer, while Rosenbusch, Zirkel, and others, have observed plates and microliths of hornblende in Scandinavian and American elæolites. To their presence the colour and peculiar lustre of the mineral have been attributed, but Rosenbusch regards the latter character as due to special molecular structure. Elæolite is a constituent of the rocks zircon-syenite, foyaite, miascite, and ditroite.

Cancrinite is probably an altered condition of nepheline. Plates of hornblende, similar to those in elæolite, have been observed in this mineral. It sometimes contains a colourless alteration-product which, under the microscope, exhibits aggregate polarisation. It is a component of the rock named ditroite in which it occurs associated with sodalite, elæolite, orthoclase, &c.

LEUCITE (AMPHIGENE).

Leucite, until within the last few years, has been believed to crystallise in the cubic system, its form closely resembling that of analcime. In 1872, however, G. vom Rath announced[2] that it crystallised not in the cubic but in the tetragonal system. His conclusions were based upon the occurrence of striæ which denote planes of twinning and which, when they extend to the edge of a face, pass across

---

[1] A. von Lasaulx, *Elemente der Petrographie*, p. 71.
[2] *Report of British Association* (Brighton), 1872, p. 79.

the edge and are continued on the adjacent face without any deviation from the direction of striation on the former face. If the leucite crystal, therefore, be regarded as a regular icosi-tetrahedron, as formerly supposed, this plane of striation would cut off the symmetric solid angles of the crystal, and the plane of twinning must consequently be regarded as parallel to a face of the rhombic dodecahedron. Twinning parallel to this face is, however, incompatible with cubic crystallisation, and leucite crystals cannot, therefore, belong to the cubic system. Following out this deduction by careful measurements, Vom Rath arrived at the conclusion that the form assumed by leucite is a combination of a di-tetragonal pyramid $4P_2$ with a tetragonal pyramid P, as indicated in the accompanying figure 58. By this means the supposed anomalous optical characters of the mineral are accounted for, and the lamellæ seen under polarised light in microscopic sections may now be safely regarded as twin lamellæ. They lie in planes parallel to the face $2P\infty$ and appear sometimes as broad, sometimes as excessively narrow, bands of bluish grey or neutral tints, differing in intensity (fig. 59), and no strong chromatic effects are ever produced by their polarisation. In such polysynthetic crystals the twinning planes lie in four directions. Although this twin structure is almost invariably present and well defined in moderately large crystals, the minute or purely microscopic crystals of leucite frequently exhibit no trace of it, as in the little leucite crystals of the sperone or leucitophyr which occurs near Rome. Were it not for

FIG. 58.

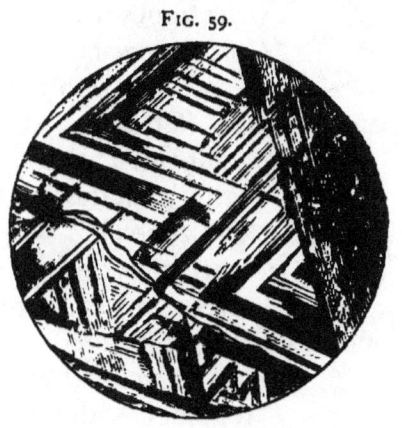

FIG. 59.

the inclusions, and the symmetrical disposition which these inclusions usually assume, such diminutive crystals of leucite might easily be overlooked, or mistaken for some other mineral; and, if this fact be borne in mind, it is quite possible that leucite may yet be detected in rocks in which its existence has never been suspected. According to the present state of our knowledge, it is a mineral of very restricted occurrence, being mainly confined to a few comparatively limited areas, such as the neighbourhood of Vesuvius, Rome, the Eifel, Saxony, Bohemia, the island of Bawian, north of Java, and in Wyoming, U.S. Sections of very minute crystals of leucite frequently show ill-defined boundaries, so that their appearance is more that of a rounded granule than of a properly developed crystal. Leucite generally affords well-defined eight-sided sections which are mostly clear, transparent, and always colourless. They nearly always contain inclosures of either colourless or brownish glass, dark, opaque granules of magnetite, microliths of felspar and augite, and sometimes granules of the latter mineral. These inclosures are generally disposed in a symmetrical manner in zones which lie parallel to the boundaries of the crystal, but the zones often fail to coincide precisely with the boundary, and they then appear as circles (fig. 60). At other times they are differently arranged as shown in fig. 61. Frequently microliths, granules, and glass pores are all present in one zone, but they are often arranged in a systematic manner, the microliths and glass-inclosures alternating with one another. Sometimes these inclosures are congregated in the centre of the crystal either in symmetrical groupings (fig. 62), or else huddled together without any definite arrangement; at other times they are scattered promiscuously throughout the crystal. Occasionally, but rarely, minute rods of glass occur in leucite crystals; sometimes they are short and thick with rounded ends, and

FIG. 60.

FIG. 61.

FIG. 62.

they are often incrusted with granular matter (figs. 63 a and 63 c). Occasionally extremely delicate rods are visible which traverse the leucite for long distances in perfectly straight lines, then end abruptly in nodes of granular matter (fig. 63 b),

FIG. 63.

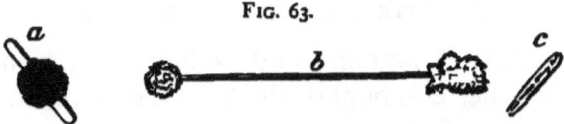

and often appear to be continued in another direction in a series of fine, straight rods, thus forming a zig-zag arrangement, nodes of granular material marking the points where they divaricate.[1]

SCAPOLITE (PARANTHINE, WERNERITE).

This mineral occasionally occurs in granite and in some metamorphosed limestones. Its chemical composition is represented by the formula $2Al_2O_3, 3SiO_2 + 3(RO, SiO_2)$, in which RO mostly signifies K_2O, but generally Na_2O is also present, and at times traces of CaO are met with. Scapolite crystallises in the tetragonal system, the common forms being the tetragonal and di-tetragonal prisms terminated by pyramids. It also occurs massive and granular. Before the blowpipe it intumesces and fuses to a white glass. It is only imperfectly soluble in hydrochloric acid. Its colour is white, bluish, greenish, and occasionally reddish. It is sometimes transparent, but more frequently opaque. It cleaves parallel to the faces of the prisms. The cleavage is more perfect parallel to $\infty P \infty$ than in the planes parallel to ∞P. The hardness is about 5·5.

Scapolite has strong double refraction. Transverse sections of scapolite crystals exhibit well-marked fissures parallel to the cleavage $\infty P \infty$. Scapolite rock is a granular massive aggregate of the mineral together with orthoclase. Under the microscope scapolite is usually seen to contain

[1] *Trans. Royal Mic. Soc.*, vol. xv. p. 180.

greenish alteration-products which sometimes have a fibrous structure at right angles to the cleavage planes.

The species meionite and marialite are closely related to scapolite.

SODALITE, HAUYNE, AND NOSEAN

are all silicates of alumina and soda, and each in addition contains another compound which serves to characterise the several species; thus,

$$\left.\begin{array}{lll}\text{Sodalite contains} & 2\text{NaCl} \\ \text{Hauyne} \quad \text{,,} & 2\text{CaSO}_4 \\ \text{Nosean} \quad \text{,,} & \text{Na}_2\text{SO}_4\end{array}\right\} + 3(\text{Na}_2\text{O}, \text{SiO}_2 + \text{Al}_2\text{O}_3\ \text{SiO}_2).$$

All these three minerals crystallise in the cubic system, the usual form being the rhombic dodecahedron. Since they all crystallise in the cubic system they are all singly refractive. They all cleave parallel to faces of the rhombic dodecahedron.

Before the blowpipe sodalite fuses with intumescence to a colourless glass; when fused with microcosmic salt and oxide of copper it gives the blue colouration to the flame, characteristic of the presence of chlorine. Hauyne and nosean when fused with carbonate of soda yield the usual reaction for sulphuric acid.

All three of these minerals are decomposed by acids with separation of gelatinous silica. In the case of hauyne sulphuretted hydrogen is evolved when the mineral is treated with sulphuric acid, but nosean under similar circumstances evolves none.

Under the microscope, sections of nosean are four-sided or six-sided, but the outlines are often irregular. The crystals, when fresh and unaltered, appear clear and colourless in the interior of the sections, but are bordered by a somewhat broad and dark external band which coincides with the boundaries of the section; occasionally crystals exhibit a concentric series of these bands. The junction of this dark border with the clear internal portion of the crystal is rather sharply defined (fig. 64). Sometimes a clear narrow outer zone surrounds the dark border, and a dark spot is often to be seen in the very centre of the crystal. The clearer portions show, under a tolerably strong magnifying power, variable quantities of fine dark dust, and dark but very fine striæ which cross one another at angles of 60°, 90°, and 120°. Under still higher powers the striæ are seen to consist of rows of minute glass inclusures and opaque granules.

FIG. 64.

The dark borders and zones of the nosean crystals are also seen to be composed of these inclusures very closely aggregated, while, at times, whole crystals appear nearly opaque owing to the great quantity of these inclusures and the multiplicity of the striæ formed by them. It may here be suggested that the clear spaces in nosean crystals are due to the segregation of these dusty inclusures towards boundaries which represent successive zones of accretion, and that this segregation implies the consequent abstraction of the dust from those portions which ultimately appear clear; a proceeding analogous to that which seems to have taken place in some tachylytes as observed both by the late Herman Vogelsang[1] and by the author.[2] Acicular microliths, opaque granules and glass, and fluid lacunæ are the principal inclusures met with in nosean. When the mineral undergoes alteration it becomes yellowish and more or less obscure, a fibrous and at times radiate structure supervenes, and the whole crystal passes into zeolitic matter which polarises in variegated colours. Nosean occurs in some phonolites, basalts, leucitophyrs, nephelinites, &c.

The sections afforded by crystals of hauyne are, like

---

[1] *Die Krystalliten*, p. 112. Bonn, 1875.
[2] 'On the Microscopic structure of Tachylyte from Slievenalargy, Co. Down,' F. Rutley, with an analysis by Dr. S. Haughton. *Proc. Royal Irish. Geol. Soc.*, 1877.

those of nosean, six-sided or four-sided. The microscopic structure of the two species shows in many cases no appreciable difference, but it is not so constant in character in hauyne as it is in nosean. Sometimes crystals of hauyne contain scarcely any inclosures.[1] Hauyne usually has a stronger blue colour than nosean, especially in the darker zones, but a blue colouration may be imparted to nosean by heating the mineral as shown by Dressel :[2] the phenomenon being due to the change which the contained sulphide of sodium undergoes, and in the rocks of the Laacher See in the Eifel, which contain nosean, hauyne is found as a substitute in those ejected blocks and lapilli which bear traces of fusion. These facts, as pointed out by Dr. A. von Lasaulx, of Breslau, indicate the probability that hauyne and nosean are mere varieties of the same mineral species.

The red colour of some hauyne is due to the presence of scales of peroxide of iron which are regarded by Zirkel as of secondary origin.[3] Crystals of hauyne, like those of nosean, contain minute glass inclosures and dusty matter, which constitute fine, microscopic striæ which follow the directions of the crystallographic axes. Both blue, bluish-grey, and colourless matter occur in some crystals of hauyne, the different colours occasionally marking definite zones, but more usually forming irregular flecks and patches of colour which bear no relation to crystallographic form; the boundary between the darker blue and the lighter colourless matter is, however, ill defined, and never shows a sharp line of demarcation.

The crystals of hauyne which occur in the basalts of the Laacher See, usually have a broad dark border which shades off towards the interior of the crystal into a bluish-grey and

---

[1] *Elemente der Petrographie,* A. von Lasaulx, p. 73. Bonn, 1875.
[2] *Neues Jahrbuch für Mineralogie,* p. 565, 1870.
[3] *Mikroskopische Besch. der Min. u. Gesteine,* F. Zirkel, p. 163. Leipzig, 1873.

Sodalite. 115

more or less translucent matter (fig. 65), but in some very small crystals no clear area is visible in the centre, even in excessively thin sections, the whole crystal being nearly or quite opaque. The striæ in the quadrangular sections cross one another at right angles. Crystals of hauyne from other localities possess a clear border with sharply-defined internal and external boundaries, the inner portion of the crystal being crossed by striæ which sometimes intersect at right angles (fig. 68), and at others follow three directions, each set of striæ forming a series of parallel lines which run either at right angles to the opposite faces of the crystal (fig. 67), or pass in directions which would correspond with lines drawn between opposite angles of the six-sided section (fig. 66). In both cases the striæ do not intersect, but are divided either by dark or clear lines which radiate from the centre of the crystal; in the former case joining opposite angles, and in the latter passing from the centre to the middle of the faces. The striæ are found under high amplification to consist of opaque granules, gas pores, and minute glass inclosures.

FIG. 65.

FIG. 66.

FIG. 67.

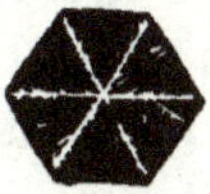

FIG. 68.

Sodalite occurs in rocks either in an uncrystallised condition, or in crystals which yield six-sided or else quadrangular sections; the latter are very frequently distorted, so that the alternate or opposite sides of the section are unequal. They usually present a yellowish, grey, or blue colour. The rhombic-dodecahedral cleavage is represented, under the microscope, by undulating cracks. The crystals sometimes appear to be remarkably pure, at others they are crowded with various inclosures; the most common being steam pores. A singly-refractive substance containing fixed bubbles also occurs in some sodalite and this singly-refracting matter, which frequently has a dark border, forms well defined rhombic dodecahedra which contain them; they do not, however, contain any bubbles. Fluid lacunæ which change on the application of heat are found sparingly in the sodalite of Somma (Vesuvius), while complete included crystals often contain inclosures of nepheline, augite, meionite, and biotite.

Olivine

is a common constituent of many eruptive rocks, in which it occurs sometimes in the form of crystals whose angles frequently appear more or less rounded, and sometimes as rounded granules which in some cases form rounded aggregates, occasionally showing traces of crystal faces. The faces most commonly presented by crystals of olivine are those of the rhombic prism ∞ P (giving an angle of little over 130°), the rhombic pyramid P, the macropinakoid $\infty \bar{P} \infty$, the brachypinakoid $\infty \breve{P} \infty$, the macro- and brachydomes $\bar{P} \infty$ and $\breve{P} \infty$ and the basal plane o P. Olivine has two directions of cleavage, one parallel to the macro- the other to the brachypinakoid, the former being very imperfect and the latter rather distinct. Olivine varies in hardness from 6·5 to 7. Its fracture is conchoidal and its colour is not only olive or bottle-green, but at times brownish and yellowish. Where, therefore, olivine occurs in rounded imbedded granules, and displays neither its characteristic form of crystal nor its common green colour, its hardness and its conchoidal fracture cause it greatly to resemble quartz, which also occurs in rounded granules in certain rocks of eruptive origin, but a mistake of this kind is only likely to occur in the hasty examination of hand-specimens in the field. The other minerals, however, with which the granules are associated usually give some clue to their probable nature. Powdered olivine is easily decomposed by hot hydrochloric acid or by sulphuric acid; separation of gelatinous silica occurring in either case. Before the blowpipe alone, only the highly ferruginous varieties are fusible, forming a black magnetic globule; but with borax the ordinary olivine may be fused to a clear green bead. The foregoing reactions certainly serve to distinguish olivine from quartz, but the microscopic characters of the two minerals are sufficiently distinct to prevent any chance of mistake. The general formula of olivine is $(MgO, FeO)_2 SiO_2$. It sometimes contains some lime, alumina, and protoxide of manganese, traces of titanic, phosphoric, and chromic acids, and the protoxides of nickel and cobalt. Potash, soda, and small quantities of water have also been detected in olivine. Olivine frequently occurs in meteorites, often forming a large proportion of them. In some of these meteoric olivines traces of arsenious acid, fluorine and oxide of tin have been discovered. The alterations which olivine undergoes are either due to peroxidation of the iron, or to its removal by water charged with carbonic acid, in which case some of the magnesia may also be removed. Changes of the former class cause the mineral to assume a reddish or brownish colour, and sometimes render it iridescent. This process when further advanced sets up a micaceous structure in the mineral,[1] and Dana mentions an instance in which a basalt, owing to this change in the olivine, was represented to be a mica slate. Changes of the latter class, viz. by water charged with carbonic acid, give rise to the formation of serpentine, steatite, &c.

Under the microscope olivine appears almost colourless when in very thin sections, and of a light greenish tint in those of moderate thinness. Its dichroism is very feeble in thick, and scarcely perceptible in thin sections. It is doubly refractive, polarising when in a fresh, undecomposed state, in moderately strong colours, but these are much more feeble than those displayed by quartz. The axes of elasticity coincide with the crystallographic axes, and corresponding extinctions may be observed by rotating the section between crossed Nicols. The surfaces of sec-

---

[1] Dana, *System of Mineralogy*, 5th edition, p. 258.

tions of olivine are nearly always rough, since the ordinary grinding is never capable of imparting a smooth polished face to the section, and these roughened surfaces, which, when examined under the microscope, present an appearance somewhat like that of ground glass, are clearly perceptible in sections mounted in Canada balsam and covered in the usual way.

Lines of accretion, such as those which form zones corresponding with the boundaries of sections of augites and felspars, have not yet been observed in sections of olivine, even when they occur in rocks in conjunction with zoned crystals of those minerals. Olivine does not generally occur in the form of very minute crystals or microliths, the crystals and granules being usually sufficiently large to be distinguished without the assistance of a lens.

Crystals of olivine generally afford six- or eight-sided sections, the angles of which are often rounded, a fact regarded by some observers as indicative of a secondary fusion. Quadrangular sections are not uncommon, and unsymmetrical forms frequently result from obliquely-cut crystals. Olivine is often traversed by strong and irregular fissures which bear no relation to the form of the crystals, and somewhat resemble the fissures visible in suddenly cooled glass. The alteration of olivine into serpentine commences along these fissures and also along the boundaries of the crystal. It appears under the microscope as a finely fibrous green fringe, the fibres lying at right angles to the surfaces from which they originate. As the alteration proceeds, this fibrous structure extends further inwards until the whole crystal is converted into a mass of interlacing and contorted, or radially disposed, fibres, and no longer displays any of the optical characters which formerly belonged to it. Gas pores, fluid lacunæ, consisting generally of liquid carbonic acid, glass lacunæ and crystals, granules and microliths of magnetite are the most common inclosures which occur in olivine. The gas pores are often arranged in rows. The microliths frequently assume peculiar forms, being sometimes zig-zag, sometimes claviform or acicular. Inclosures of augite, felspar, leucite, &c., are never met with in olivine.

Hypersthene.

Crystalline system rhombic. Usually occurs, in rocks, in a crystalline condition or in granules, but seldom in actual crystals. The mineral shows a strongly marked cleavage parallel to the brachypinakoid, very imperfect cleavage parallel to the macropinakoid, and a tolerably perfect prismatic cleavage. The colour is black, greyish, brownish-green, or pinchbeck-brown, and presents in certain directions a glistening appearance or bronze-like lustre. The chemical composition is $(MgO, FeO) SiO_2$.

Before the blowpipe it fuses to a black enamel, and, on charcoal, to a magnetic mass.

When tested with a single Nicol, thin sections of hypersthene appear strongly pleochroic. The colour is mostly greyish-green in the direction of the principal axis; in the direction of the macrodiagonal it is reddish-yellow, and in that of the brachydiagonal it is hyacinth-red. Hypersthene is not, however, so strongly pleochroic as hornblende, but it possesses this character in so marked a degree that it at once serves to distinguish it from diallage, bronzite, and enstatite. The hypersthene of Labrador shows numerous imbedded lamellæ which were regarded by Vogelsang as possibly diallage, but A. von Lasaulx thinks this improbable, since similar imbedded lamellæ have been observed in diallage from several localities. These plates have been referred by different authors sometimes to göthite, to specular iron, to brookite, &c. Their true nature is, however, as yet undetermined. Minute granules of magnetite are of frequent occurrence in hypersthene, often following one another in definite directions parallel to the principal axis. Bořicky mentions the occurrence of minute plates of calcite imbedded in the direction of principal cleavage in the hypersthene of a rock from Tellnitzthal.[1]

ENSTATITE.

Crystalline system rhombic. Cleavage parallel to the faces of the rhombic prism ∞ P. The cleavage planes make by their intersection angles, according to Descloizeaux, of 88° and 92°; according to Kengott 87° and 93°. Enstatite also affords two other directions of less perfect cleavage, one parallel to the macro-, the other parallel to the brachy-pinakoid. The cleavage surfaces sometimes have a fibrous appearance, and usually a rather pearly or vitreous lustre, while in some varieties it is metalloidal. In colour the mineral ranges from greyish- or yellowish-white to green and brown. It occurs massive and lamellar as well as crystallised. It is almost infusible before the blowpipe, the edges of only very thin splinters becoming rounded. It is insoluble in hydrochloric acid. Enstatite becomes altered to schiller-spar or bastite, talc, &c.

The formula of enstatite is $(MgO, FeO) SiO_2$.

Enstatite is strongly dichroic, the greatest colour differences being clear brownish-red and pale sea-green.

Enstatite occurs in lherzolite and certain gabbros, and, according to Professor Maskelyne, some of the diamond-bearing rocks of South Africa contain more or less of this mineral.

BRONZITE.

This mineral crystallises in the rhombic system. It is very difficultly fusible, the edges only of very thin splinters becoming rounded before the blowpipe. It is insoluble in acids. In its microscopic structure it is more closely allied to hypersthene than to enstatite. It resembles the latter in its strongly fibrous structure, and the former in the character of its numerous inclosures which occur in the form of lamellæ, or, as in the Kupferberg bronzite, of 'elongated

---

[1] *Petrographische Studien an den Basalt gesteinen Böhmens*, von Dr. E. Boŕicky, p. 16. Prag, 1874.

stripes of an acicular character set perfectly parallel to the fibrous structure and varying in colour from dark honey yellow to deep brown.'[1]

Some bronzite is very feebly dichroic, or scarcely dichroic at all. In this respect bronzite differs both from enstatite and hypersthene. The directions of cleavage in bronzite are, parallel to the brachypinakoid, highly perfect; parallel to the right rhombic prism, imperfect; parallel to the macropinakoid, most imperfect. The formula of bronzite is (MgO, FeO) SiO_2. Some analyses show the presence of lime and alumina.

PYROXENE GROUP.

All the minerals of this group crystallise in the monoclinic system and are bi-silicates of different protoxides, having the general formula RO, SiO_2, the protoxides often being converted into sesquioxides. The protoxides are lime, magnesia, sometimes potash and soda, and the protoxides of iron and manganese. The different members of this group vary more or less in composition, a fact due to the respectively isomorphous character of their protoxides and sesquioxides. The most important difference in the chemical composition of the pyroxenes and the amphiboles, both of which groups have the same general formula, lies in the fact that in the former lime is in all the varieties of the group an important constituent, while in some of the varieties of the latter group it is either totally absent or occurs only in very small quantity.

Chemically, the pyroxenes may be divided into those which are poor in, and those containing from three to over nine per cent. of, alumina. The diallages in this respect range between the two divisions. Some of the so-called diallages (metalloidal diallage), belong rather to enstatite than to pyroxene, since the crystallisation is rhombic. Descloiseaux regards the diallage occurring in the Cornish serpentines as probably a form of enstatite.[2]

[1] *Mik. Beschaff. d. Min. u. Gest.*, Zirkel, p. 187.
[2] *System of Mineralogy*, J. D. Dana, 5th edition, 1871, p. 209. In

Augite crystals frequently occur twinned, the plane of twinning being parallel to the orthopinakoid.

The angle formed by the oblique rhombic prism in augite is 87° 5'. The corresponding angle in hornblende crystals is 124° 30'. This great discrepancy in their angular measurements serves to distinguish the minerals of the one species from those of the other, and even under the microscope, when the planes of sections often lie obliquely to the principal crystallographic axis, a rudely approximate measurement frequently enables the observer to discriminate between hornblende and augite. Another means of distinguishing between these two minerals is afforded by the strong dichroism which usually characterises hornblende, and the very feeble dichroism presented by sections of augite under the microscope; the latter giving only a succession of slightly different tints, while in hornblende an actual difference in colour is to be observed.

It sometimes happens, however, that very thin sections of hornblende exhibit only feeble dichroism, and that thick sections of augite may show this character in a rather marked manner. In augite the axis of least elasticity (c) makes an angle of 38°—39° with the vertical crystallographic axis, while in hornblende it makes a corresponding angle of 1°—18°.

Such dichroism as augite sections may exhibit is always, according to Mr. Allport's observations,[1] connected with a purple tinge, and he has noted this fact in augites occurring in rocks of very different ages and derived from widely separated localities. He also states that dichroism is more strongly elicited from those augite crystals which by ordinary illumination exhibit a variation in colour, *i.e.* when one end or side of a crystal appears purplish-brown and

Descloiseaux's *Manuel de Minéralogie*, t. i., 1862, there is, however, no mention of this; the Cornish serpentines being there, and at that time, cited as containing diallage.

[1] 'On the Microscopic Structure and Composition of British Carboniferous Dolerites,' S. Allport, *Q. J. G. S.*, vol. xxx. p. 536.

shades off into yellowish-brown towards the opposite end or side.

By ordinary transmitted illumination thin sections of augite appear of a greenish or yellowish-brown colour. Wedding has noted the occurrence of greenish and brownish layers, sharply defined, in the augites of the Vesuvian lavas, and concludes that the latter colour is not due to oxidation of ferrous compounds constituting the greenish portions of the crystals.

Similar bands, frequently well marked and of variegated colour, are often to be seen by polarised light in sections of augite and represent twin lamellæ which lie parallel to the orthopinakoid. Even by ordinary transmitted light they may sometimes be recognised as exceedingly delicate striæ.

Sections of hornblende and augite, when cut at right angles to the principal axis, may usually be distinguished by the augite giving eight-sided sections, as a rule; while those of hornblende are generally six-sided.

In some sections of augite crystals striæ are visible under the microscope, which correspond with the boundaries of the section, or, in other words, with the external form of the crystal. These lines lie one within the other, constituting zones of variable width, and are often rendered still more apparent by granules of magnetite, microliths, and small cavities, which follow or coincide with these lines in a remarkably regular manner. In polarised light the concentric bands exhibit different tints and appear more strongly marked. They no doubt represent lines of accretion.

The inclosures in augite crystals consist, so far as they have yet been observed, of acicular microliths of augite itself. Triclinic felspars and leucite also occur in minute crystals imbedded in augite, and the sides of augite crystals are often penetrated by crystals of apatite. Magnetic and titaniferous iron are of frequent occurrence in augite, sometimes almost entirely replacing it. Furthermore, augite often contains inclosures of amorphous glass, sometimes spherical, sometimes irregular in form, and occasionally cavities containing fluid have been observed. In some exceptional cases the augites of certain basalts are merely represented by a thin line or border of augitic substance, the interior of the crystal being filled with the admixture of minerals which constitutes the ground-mass of the rock. The nature of the minerals, &c., inclosed in augite crystals depends of course very materially upon the mineral composition of the rocks in which the augites occur.

This mineral does not merely occur in rocks in the form of well-defined crystals, but also as acicular microliths which are generally greenish or yellowish-brown, unless they are exceedingly minute, in which case they are almost colourless.

These microliths vary considerably in form, some being straight, some curved, while occasionally they are forked at the ends, or assume club-like forms. Sometimes they lie independently scattered through the matrix of the rock, at others they are clustered together, especially around the margins of crystals.

Diallage, which is a common constituent of some rocks, is usually regarded as a variety of augite. It exhibits a highly perfect cleavage parallel to the orthodiagonal, thus differing from augite, in which the corresponding cleavage is imperfect. Both of these minerals also possess a perfect, or tolerably perfect, cleavage parallel to the faces of the oblique rhombic prism (Pl. II. p. 173).

Diallage resembles augite in displaying very weak dichroism, so weak that it gives rise to no marked difference of colour, and may thus be distinguished from hornblende and hypersthene. Thin plates of diallage when examined microscopically are often seen to include numerous little tabular crystals and acicular microliths. The tabular crystals mostly run in definite planes parallel to the ortho- and to the clinopinakoids of the diallage and also to a third plane which lies obliquely to that of the orthopinakoid. 'Sections taken parallel either to the ortho- or clinopinakoid always exhibit the broad sides of those little imbedded crystals which are conformable to the face under examination, together with the line-like transverse sections of similar crystals which lie with their broad sides parallel to the other face.'[1] The acicular microliths in some diallages also follow definite directions and appear to cross one another obliquely, forming a somewhat lozenge-shaped net work. Fluid lacunæ have likewise been observed in the diallage of a Silesian gabbro. Diallage sections usually exhibit a pale greenish or brownish-yellow tint by ordinary transmitted light.

Asbestus is in some cases a fibrous form of augite, but most asbestus is hornblendic in its affinities.

Breislackite is a capillary or woolly form of pyroxene, occurring in the lavas of Vesuvius and Capo Di Bove. Although definitely known to be a pyroxene, the precise species to which it belongs has not yet been determined.

Diallagic Augite.—A form of pyroxene intermediate in character between augite and diallage has been noticed by Professor E. Bořicky, of the university of Prag, in his 'Petrographische Studien an den Melaphyr Gesteinen Böhmens,' p. 19. He describes it as an augite whose sections may be distinguished from ordinary augite by the occurrence of straight and parallel fissures or striæ which, in longitudinal sections of the crystals, cross the coarser cleavage planes at angles of from 70° to 90°. Professor Bořicky considers these lines to represent the edges of sections of delicate lamellæ which he regards as twin-like intergrowths, most probably lying parallel to the basal planes of the crystals. He states that in the melaphyres of Neudorf near Lomnitz the diallagic augite sections are broad, irregulalry bounded, and contain numerous bubbles and stone lacunæ or 'stone cavities' (*Schlackenkörnchen*) which are often ranged in lines, either parallel to the fissures or actually along them. Such fissures or striæ often occur in one part only of an individual crystal. The mineral is not dichroic, and polarises in strong colours, the crystal sections sometimes presenting

---

[1] *Mikroskop. Beschaff. der Min. und Gest.*, Zirkel, p. 181.

iris-coloured margins. He has noted their occurrence in the melaphyres of Hořensko, Lomnitz, Neudorf near Lomnitz, and Zdiretz in Bohemia.

Altered Conditions of Pyroxene.

It is important that the microscopist should have some knowledge of the alterations which pyroxene undergoes, in order rightly to understand the nature of the pseudomorphs after the different members of this group which so frequently occur in eruptive rocks.

The simplest kind of alteration which pyroxenic minerals experience is hydration. It is frequently accompanied by a loss of silica. The lime and iron contained in these minerals also undergoes considerable diminution, or is totally removed by water charged with carbonic acid or holding carbonates in solution.

The following are some of the results of the alteration of pyroxene.

Table showing the Approximate Chemical Composition of the Various Minerals which Result from the Alteration of Pyroxene.

	Si	Äl	Čr	Ḟe	Fe	Mn	Mg	Ca	Na	K	Ti	H	
Augite	51	3	—	—	6	3	13	24	—	—	—	—	Rammelsberg
Pycrophyll	50	2	—	—	6	—	31	1	—	—	—	10	Rose
Pyrallolite	49½	½	—	—	1½	1	24½	10½	—	—	—	12½	Runeberg
Schiller-spar	43	2	2½	—	11	½	26	2½	—	—	—	12½	Köhler
Epidote	46	5½	—	13	8	—	12½	9	—	—	—	5	Streng
Mica	43	15	—	—	23½	—	10½	1	1	5	—	—	Kjerulf
Uralite	49	1	—	—	25½	—	12	11½	—	—	—	1	Rath
Glauconite	51½	7	—	—	21	—	6	—	2	6	—	6½	Delesse
Serpentine	41	2	—	—	2	—	42	—	—	—	—	13	Scheerer
Steatite	62½	—	—	—	1½	—	31	—	—	—	—	5	Richter
"	64	1	—	—	—	—	28	—	—	—	—	7	Tengström
Palagonite	42	12½	—	16	—	—	7	7	2	—	—	12½	Waltershausen
Hematite	—	—	—	100	—	—	—	—	—	—	—	—	—
Limonite	2½	—	—	80½	—	—	—	—	—	—	—	16	Ulmann
Magnetite	—	—	—	69	31	—	—	—	—	—	—	—	—
Titaniferous Magnetite	—	—	—	22	51½	2	—	—	—	—	24¼	—	Knop
" "	—	—	—	68	30	—	—	—	—	2	—	—	Michaelson

The list is headed by a typical analysis of unaltered augite from Schima in Bohemia. The decimals in the original analyses have been omitted, and the fractions rather differently distributed, in order to simplify the columns for reference.

In addition to these analyses may be cited one of a highly siliceous pseudomorph after pyroxene, by Rammelsberg, consisting of 85 per cent. silica, $1\frac{1}{2}$ alumina, 2 protoxide of iron, 2 magnesia, $2\frac{1}{2}$ lime, and 6 water.

Serpentine, steatite, and limonite are probably the most common of these alteration-products in British eruptive rocks. Epidote seldom gives direct evidence of its derivation from pyroxenic minerals, since it generally occurs in characteristic crystals or radiating tufts along minute lines of fracture, and not in pseudomorphs bounded by the outlines of pyroxenic forms. Nevertheless, it does not follow that it has not, in these cases, resulted from the decomposition of pyroxene. Pseudomorphs after pyroxene of quartz, opal, and calcite are also of occasional occurrence, but these are more probably due to subsequent infiltrations, than to a partial removal of the basic constituents, or, in the case of opal, to the hydration of a residuum of silica. The usual mode of deposition of hydrous silica, opal hyalite, &c., in vesicles and cavities, and on the surfaces of joint planes in eruptive rocks, tends to support this view.[1]

AMPHIBOLE GROUP.

The minerals of this group closely resemble pyroxene in chemical composition, while they also crystallise in the same system (monoclinic). They differ, however, as already pointed out, in the angular measurements of the oblique rhombic prism, which in hornblende is 124° 30′ and in augite from 87° 5′ up to 92° 55′. They are all bi-silicates

[1] According to Mitscherlich, Berthier, and G. Rose, tremolite and actinolite (both varieties of amphibole) yield, when fused in a porcelain furnace, forms similar to those of pyroxene. Crystals of augite are of common occurrence in blast furnace slags, sometimes even associated with hornblende crystals, but the latter are less frequently met with; and it is stated that Mitscherlich and Berthier, although able to produce artificial crystals of augite, failed to procure any of hornblende.

of protoxides and sesquioxides, the former being lime, magnesia, soda, potash, and the protoxides of iron and manganese, while the latter are represented by alumina and the peroxides of iron and manganese. Crystals of amphibole differ from those of pyroxene, not merely in the angular measurements of their oblique rhombic prisms, but also in the angles at which their cleavage planes intersect. In both groups the relation of the cleavages to the respective faces of the crystals is the same, but they differ in their respective facilities; the cleavage parallel to the faces of the oblique rhombic prism in hornblende being more perfect and more strongly marked than the corresponding cleavage in augite, while the cleavages parallel to the pinakoids are on the other hand less strongly marked in hornblende than in augite. Furthermore, the discrepancy in the angles of the two respective oblique rhombic prisms begets a corresponding discrepancy in the angles at which the prismatic cleavages of the two different species intersect. This circumstance is of considerable value to the mineralogist, since it is often difficult or impossible to measure the angles of the actual crystallographic faces, but it is generally possible to measure the angles of cleavage, and since these cleavages respectively coincide with the plane ∞P, the results deduced from cleavage are as good as those derived from the actual faces of the crystals.

The crystals of minerals belonging to the amphibole group usually exhibit a fine longitudinal striation.

The very feeble dichroism of augite and the strong dichroism of hornblende has already been mentioned, and, although exceptional cases may occur, it must nevertheless be regarded as a most valuable test, especially in the microscopic examination of rocks which contain minerals pertaining to one or other of these groups.

The amphiboles, like the pyroxenes, may be divided into two sub-groups, viz., those which contain little or no alumina, and those which are rich in that base (sometimes containing over 15 per cent.) The similarity in the chemical composition of amphibole and pyroxene begets a similarity in the minerals which result from their alteration, so that the alteration-products tabulated at the end of the pyroxene section may be taken as fairly representative of the changes which amphibole undergoes.

In the microscopic examination of thin sections of hornblende, the forms are often so irregular, that it is difficult to arrive at any sound deductions from their contours.

Colour also affords no safe means of discriminating between pyroxene and amphibole, since the members of both groups exhibit greenish and brownish tints. The augites and hornblendes which occur in basalt are mostly brownish in colour. The hornblende in syenite is also generally brown, but that which occurs in phonolite is mostly of a greenish tint while the augite in leucite-lavas is, as a rule, also green.

The most important microscopic characteristics of common hornblende may be summed up in the following manner. Transverse sections of crystals show two sharply defined sets of cleavage planes, each set corresponding with the opposite and alternate faces of the oblique rhombic prism and intersecting at an angle of 124° 30′ (when the section is at right angles to the principal crystallographic axis). In sections taken parallel to the chief axis of the crystal a fine longitudinal striation, indicative of a fibrous structure, is often to be observed.

Furthermore, the dichroism of hornblende is very strong, although the clear green varieties, as pointed out by Zirkel, show only very feeble dichroism and might be mistaken for augite.[1]

Some of the small crystals, such as those often occurring in phonolites, appear to be made up of fine parallel rods or elongated microliths, the margins being frequently very irregular.

Microliths of hornblende are of common occurrence in

[1] Zirkel, *Mik. Beschaff. d. Min. u. Gesteine*, p. 169.

eruptive and metamorphosed rocks; when very minute they are often colourless. They seldom present any recognisable crystalline form, but frequently exhibit longitudinal fibrous structure, often appearing to be frayed out at the ends.

Zone-like bands of accretion, similar to those which are sometimes visible in sections of augite, also occur in hornblende sections. Magnetite, apatite, nepheline, biotite, quartz, &c., occur, enclosed in hornblende crystals; glass lacunæ are also frequently met with in the hornblendes of trachytes, phonolites, pitchstones, &c.

The hornblendes of syenites, diorites, &c., contain as a rule fewer microscopic inclosures than those of volcanic rocks, such as trachytes, phonolites, basalts, &c.

Crystals of hornblende frequently exhibit a dark granular border of magnetite, and at times the latter mineral has been observed greatly to dominate over the hornblende, the hornblendic matter merely appearing as little interstitial specks between the magnetite granules. In such crystals, one may almost say pseudomorphs, the outlines of the sections still clearly denote the form of hornblende; and it seems, as suggested by Zirkel, that in these cases even a very small proportion of hornblende, in spite of an almost overwhelming percentage of magnetite, is capable of asserting its crystalline form, just as in the well known Fontainebleau sandstone a trivial proportion of calcite develops rhombohedral crystals which contain as much as 65 per cent. of sand.[1] In this latter case the sand, as stated by Delesse, was formed prior to the crystallisation of the calcite; in other words the silica and the carbonate of lime did not crystallise at the same time. It is therefore probable that the granular magnetite was also developed prior to the crystallisation of the hornblende, and was simply taken up by it.

The minerals of the amphibole group frequently show a tendency to develope long blade-like crystals. One of the

---

[1] Delesse, 'Récherches sur les Pseudomorphoses,' *Annales des Mines*, t. xvi. 1859. (Separately printed extract, p. 35.)

principal varieties, actinolite, shows this tendency in a very marked manner, the crystals arranging themselves in radiate groups. Actinolite is a magnesia-lime-iron amphibole; it is usually of a dark green colour. Under the microscope the crystals mostly appear pale green by ordinary transmitted light, and are often traversed by numerous transverse fractures, frequently accompanied by displacement of the crystal on either side of these cracks. Some of the long radiating crystals in a serpentinous rock from Cannaver Island, Galway, Ireland, display magnificent variegations of colour under polarised light.

Tremolite is a magnesia-lime amphibole $(CaO, MgO) SiO_2$. In colour it is generally white or greyish. Nephrite or jade is in part a tough compact fine-grained tremolite, having a tinge of green or blue, and breaking with a splintery fracture and glistening lustre. It usually occurs associated with talcose or magnesian rocks.'[1]

Asbestus or amianthus is a fibrous variety of pyroxene occurring in white silky fibres, which are matted together, but are easily separable. Byssolite is more compact in aggregation, the fibres are coarser as a rule and are not easily separated, the structure more resembling that of wood, while the colour is usually dark green or greenish grey. It may be regarded as an iron-manganese amphibole.

All these varieties, viz., actinolite, tremolite, jade, asbestus, and byssolite belong to the non-aluminous, or almost non-aluminous, sub-group of amphibole.

Ordinary hornblende, its variety pargasite, and smaragdite, which is a foliated grass-green form of hornblende, somewhat resembling diallage in appearance, are varieties of the aluminous sub-group of amphibole.

Hornblende is frequently twinned along a plane parallel to the orthopinakoid. This gives rise to a difference in the terminations of the crystals, one end exhibiting four faces (hemidomes), and the other two faces (basal planes).

[1] Dana, *System of Mineralogy*, 5th edition, p. 233.

Both hornblende and augite sometimes occur together in the same rock; but as a rule the former mineral is found in those rocks which contain a large percentage of silica, the associated minerals being usually quartz and orthoclase, while augite is generally found in rocks of a basic character containing triclinic felspars, and with little or no free silica. Augite and its variety, diallage, sometimes occur together in the same rock, and in such cases the petrologist often has difficulty in the precise determination of the pyroxenic constituents; the diallagoid augite[1] of Bořicky sufficiently evinces, merely from the name which has been given to it, the intermediate character which such pyroxenic minerals may sometimes possess.

Mica Group.

The minerals of this group, most commonly occurring as constituents of rocks, crystallise either in the hexagonal or in the rhombic system. They have a highly perfect cleavage parallel to the basal plane, and the thin laminæ procured by cleavage are not merely flexible but also elastic, springing back, when bent, into their original position. The hexagonal species are uniaxial, the rhombic biaxial. The micas mostly have a pearly or sub-metallic lustre. Their hardness is very variable, some of them ranging as low as 1, while others have a hardness of 3 or 4.

In chemical composition the micas vary considerably, and cannot well be represented by a general formula. They are silicates of alumina, with silicates of potash, magnesia, or lithia, and protoxides of iron and manganese. Some species are hydrous, but this condition implies alteration in most instances. It is now considered probable by some mineralogists that isomorphous mixtures of the different species may occur amongst the micas, just as they do among the felspars.

[1] *Diallagähnlicher Augit.* Bořicky, *Petrographische Studien an den Melaphyrgesteinen Böhmens,* p. 19. Prag, 1876.

RHOMBIC MICA SECTION.

Muscovite.—A potash mica, optically biaxial, crystallising in the rhombic system, and usually occurring in rhombic or six-sided tabular crystals; the lateral faces are sometimes those of pyramids, sometimes of the rhombic prism, affording in the latter case an angle of about 120°, the opposite acute angles often being truncated by faces of the brachypinakoid. In many rocks the crystals are but poorly developed, or only represented by irregularly-shaped scales, which occasionally, but rarely, exhibit a slight curvature. Cleavage basal and very perfect. Hardness = 2 to 3. Colour, mostly silvery white; seldom, but occasionally, dark brown or black.

The percentage chemical composition of muscovite may be regarded typically as:—

$SiO_2 = 46\cdot3$. $Al_2O_3 = 36\cdot8$. $K_2O = 9\cdot2$. $Fe_2O_3 = 4\cdot5$. $HF = 0\cdot7$. $H_2O = 1\cdot8$.

Before the blow-pipe it whitens and fuses on thin edges to a grey or yellow glass. Muscovite is not decomposed by sulphuric or hydrochloric acids.

Under the microscope sections of muscovite appear transparent, and exhibit clear colours. Plates, viewed at right angles to the basal plane, show tolerably strong chromatic polarisation, in this respect differing from the uniaxial micas which, under similar conditions, become dark between crossed Nicols. When tested for dichroism it shows but little change of actual colour; as a rule merely displaying a change from light to dark shades of the same colour. Two systems of striæ are often visible under the microscope on the surfaces of thin plates, the one set running parallel to ∞P and $\infty \breve{P} \infty$, the other less perfect following $\infty \breve{P}_3$ and $\infty \breve{P} \infty$. These directions of striation bear a definite relation to the optical axes. Fluid inclosures occur in some micas, but they contain no bubbles, and, as suggested by Dr. A. von Lasaulx, are doubtless merely infiltrations which have occurred along cleavage planes. Tourmaline, apatite, garnet, quartz, magnetite, and undetermined microliths have been observed as inclosures in muscovite; but, as a rule, the mineral contains few foreign matters. Newton's rings are often visible between the laminæ. Small crystals and, at times, interlaminations of biotite occur in muscovite.

Fuchsite is a variety of muscovite containing about 4 per cent. of chromic acid.

Lepidolite (lithia mica) corresponds crystallographically and physically with muscovite, for which it is frequently a substitute in granites. It usually occurs in fine scaly or granular aggregates, rather than definite crystals. The colour is generally violet, rose-red, or violet-grey, and occasionally white.

In chemical composition lepidolite may be regarded as muscovite, in which the potash is partially replaced by lithia. An analysis of lepidolite from Rozena in Moravia gave silica $= 50\cdot32$, alumina $= 28\cdot54$, peroxide of iron $= 0\cdot73$, magnesia $= 0\cdot51$, lime $= 1\cdot01$, lithia $= 0\cdot70$, fluoride of lithium $= 0\cdot70$, fluoride of potassium $= 12\cdot06$, fluoride of sodium $= 1\cdot77$, rubidia $= 0\cdot24$, cæsia $=$ traces, water $= 3\cdot12$. Lepidolites from Utö in Sweden have yielded over $5\cdot5$ per cent. of lithia. Before the blowpipe lepidolite colours the flame purple-red. After fusion before the blowpipe, it is completely decomposed by acids; but otherwise it is only imperfectly soluble.

Damourite and *Sericite* are hydrous potash micas usually occurring in scaly aggregates, but their crystallographic system has not yet been determined. Sericite occurs in undulating scales which have a fibrous structure. These wavy scales often run through the schistose rocks in which they occur in tolerably parallel directions; at other times they anastomose or form a mesh-work. The fibrous structure distinguishes it from mica. Each fibre has an individual polarisation, a character which is very constant. It may be distinguished from chlorite by the absence or feebleness of dichroism. By ordinary transmitted light thin sections of sericite are colourless, yellowish, or greenish, but never of so strong a green colour as chlorite.[1]

Paragonite is a hydrous soda mica.

Margarodite, crystallographically and optically, is almost identical with muscovite, and results from the hydration of that mineral. Chemically it approximates to damourite. The colour is generally silvery white, and it has a more pearly lustre than muscovite.

Phlogopite crystallises in the same system, and has the same cleavage as muscovite. The divergence of the optical axes is from $5°$ to $20°$. The colour is usually brown or copper-red, sometimes white. Thin laminæ often show a stellate figure when a candle flame is looked at through them; this asterismus is due to the presence of included microliths or small crystals, which follow three definite directions. The chemical composition is essentially silicate of alumina, magnesia, potash, and frequently soda. Magnesia is sometimes present from 20 to 30 per cent. It is a mica which mostly occurs in crystalline limestones, dolomites, and serpentines. The dichroism of this mineral is strong. Its microscopic character is not, however, sufficiently well marked to enable it easily to be distinguished by this means from muscovite or lepidolite, chemical analysis being the only trustworthy method of discrimination in this case.

HEXAGONAL MICA SECTION.

Biotite crystallises in the hexagonal (rhombohedral) system.[2] Colour, black or dark green. Very thin laminæ, appear brown, greenish, or red by transmitted light. Chemical composition, silicate of magnesia, potash, iron, and

---

[1] *Memoire sur les Roches dites Plutoniennes de la Belgique*, Poussin and Rénard. Bruxelles, 1876, p. 164.

[2] According to Descloizeaux some specimens of biotite are optically biaxial, but the observed divergence of the optical axes is very slight. In these cases the mineral must be regarded as rhombic in crystallisation, and closely related to, if not identical with, phlogopite.

alumina. The percentage composition of the mineral varies considerably. The basal cleavage is highly perfect, and the laminæ are flexible and elastic, as in other members of the mica group. The mineral is only slightly acted upon by hydrochloric acid, but is decomposed by sulphuric acid, leaving a residue of glistening scales of silica.

Under the microscope, crystals which lie parallel to the basal plane appear dark between crossed Nicols, but sections taken in other directions through the crystals show very strong dichroism. As observed by A. von Lasaulx, hornblende, tourmaline and epidote are the only other minerals which exhibit equally strong dichroism. Sections transverse to the basal plane show a fine striation, which represents the cleavage, and the crystals often appear frayed out at the ends.

Lepidomelane occurs in small six-sided tabular crystals, or in aggregations of minute scales. Colour, black. Lustre, adamantine or somewhat vitreous. Easily decomposed by hydrochloric acid, leaving a fine scaly residue of silica. In chemical composition it is an iron-potash mica. An analysis of lepidomelane from Carlow Co., Ireland, by Haughton, gives

$SiO_2 = 35.55$. $Al_2O_3 = 17.08$. $Fe_2O_3 = 23.70$. $FeO = 3.55$. $MnO = 1.95$. $MgO = 3.07$. $CaO = 0.61$. $Na_2O = 0.35$. $K_2O = 9.45$. $H_2O = 4.30$.

This mica occurs in some of the Irish, and, according to Allport, in some of the Cornish granites. It is also found in certain Swedish rocks. Lepidomelane is usually regarded as hexagonal, and consequently as optically uniaxial, but some crystals have been observed which appear to be biaxial, a very slight separation of the axes being discernible.

CHLORITE.

Crystalline system hexagonal. It occurs sometimes in a granular form ('Peach'), sometimes in small green crystals and scales. Its formula is $2(2RO, SiO_2) + Al_2O_3, 3H_2O$ in which RO signifies MgO and FeO. The average percentage of magnesia is about 34 and that of water over 12. It is essentially a product of the decomposition of other minerals. When heated in a glass tube it gives off water. Before the blowpipe it is fusible with difficulty on thin edges. It is decomposed by HCl and completely soluble in hot H_2SO_4. Chlorite is optically uniaxial, consequently a thin crystal of chlorite, when lying with its basal plane at right angles to the axis of vision, exibits no dichroism, but sections of chlorite crystals taken either obliquely, or at right angles to the basal plane, show feeble dichroism, giving a change from pale to deep tints of green when examined with a single Nicol, although, as a rule, no actual difference of colour is discernible. Chlorite often contains fluid lacunæ. It frequently forms fibrous and radiate aggregates. Magnetite and radiating nests of actinolite are commonly associated with chlorite. Clinochlore, which is monoclinic in crystallisation and optically biaxial, also frequently occurs in admixture with chlorite.

TALC.

Crystalline system rhombic? or, according to some authors, hexagonal. Has a highly perfect basal cleavage. Cleaved plates are flexible, but not elastic: by this character it may be distinguished from mica. Hardness = 1. Crystals are rare. Colour silvery-white to various shades of green. Lustre pearly. Unctuous to the touch. Formula $(MgO)_6 (SiO_2)_7$. Before the blowpipe it turns white and exfoliates. It is, neither before nor after ignition, soluble in either hydrochloric or sulphuric acids, thus differing from chlorite. It also differs from chlorite in showing no dichroism. Under the microscope it appears as imperfectly developed scales, with a fibrous structure: these often have frayed margins. Steatite may be regarded as a massive variety of this mineral.

TOURMALINE.

Crystalline system hexagonal (rhombohedral). The crystals commonly occur in long prismatic forms. The development is often hemihedral, thus giving rise to triangular prisms. The terminations of tourmaline crystals are frequently composed of a great number of faces, and the one termination differs from the other. The crystals, when heated and freely suspended, exhibit polar electricity, a phenomenon which usually accompanies hemimorphism. Cleavage rhombohedral but very imperfect. Crystals generally striated longitudinally. They often form radiate groups. The chemical composition of tourmaline is very variable. The general formula is $3R_2O_3 . SiO_2 + 3RO . SiO_2$. All the varieties contain silicate of alumina with about 4 to 10 per cent. of boracic acid, some contain protoxide, some peroxide, and some both protoxide and peroxide of iron. Magnesia, soda, lime, phosphoric acid, lithia, and fluorine also occur in some varieties. The fusibility of some varieties is effected with more or less difficulty before the blowpipe. Others fuse with comparative ease; in some cases the fusion is accompanied by intumescence. Heated with powdered fluor-spar and bi-sulphate of potash, all the varieties impart to the oxidising flame the green colouration indicative of boracic acid. Tourmaline is strongly doubly-refractive and strongly dichroic. In thin section, by ordinary transmitted illumination, crystals, or portions of crystals, frequently exhibit a deep blue colour: this is especially the case when the tourmaline is associated with quartz. Transverse sections of prisms often afford triangular forms. Although in its strong dichroism it resembles hornblende, the latter mineral may be distinguished from it by its well-marked cleavage-planes. Biotite is not likely to be mistaken for tourmaline owing to the lamellar structure of the former mineral: a structure almost always visible in those sections of biotite which are not cut parallel to the basal plane and which exhibit dichroism. Tourmaline contains, as a rule, few inclosures, the most common being fluid lacunæ. The black variety of tourmaline is termed schorl, and is of frequent occurrence in certain rocks, especially in granites, which generally become schorlaceous near their contact with other rocks. In such cases it is not uncommon to find the schorl segregated into nests, or small spheroidal masses.

EPIDOTE.

Crystalline system monoclinic. Cleavage parallel to the orthopinakoid perfect, parallel to the basal plane, very perfect. The cleavage planes intersect at an angle of 115° 24'. Colour usually green or yellowish-green, sometimes brown. Epidote occurs at times in a granular or massive condition; the crystals commonly form radiate or fan-shaped groups, and either form nests, or line fissures and cavities. The mineral is essentially an alteration-product, consisting of silicate of alumina, lime, and peroxide of iron, with variable amounts of oxide of manganese and water. Before the blowpipe it usually gives an iron or manganese reaction with fluxes. There are several varieties, one containing about 14 per cent. of manganese oxides.

Under the microscope epidote exhibits strong pleochroism. When tested with a single Nicol it does not, however, show this property so strongly as hornblende.

Inclosures are rare in epidote; nevertheless fluid lacunæ have been observed in it. The way in which epidote occurs in rocks forming little fan-shaped aggregates of radiating needles or fibres along fissures, &c., its bright yellowish-green colour, and its strong pleochroism—all of these characters taken in conjunction serve, as a rule, to distinguish it from other minerals. Epidote often forms little fringes around hornblende crystals. Chlorite is one of the minerals which most resembles it in its mode of occurrence. In doubtful cases, the respective strength of dichroism, the absence of dichroism in plates of chlorite viewed perpendicularly to the basal plane, the hexagonal form of such chlorite plates, and the difference in the colour of the two minerals, are points which should be looked for and taken into consideration.

Sphene (*Titanite*)

crystallises in the monoclinic system; common form, the oblique-rhombic prism. The crystals are usually thin and have sharp wedge-like edges. Colours, brownish-grey, yellow, green, and black. Transparent to opaque.

Approximate chemical composition. Silica$=30-35$, titanic acid$=33-43$, lime$=21-33$ per cent. The formula may be written :—$CaO, 2SiO_2 + CaO, 2TiO_2$.

Under the microscope, sections of sphene generally appear, by transmitted light, brownish yellow, yellow, sometimes red, reddish brown, and colourless; they show distinct though not strong, pleochroism, except when the sections are very clear and colourless or of extreme thinness. Infiltration products, consisting either of hydrous or anhydrous oxide of iron, are sometimes seen between the planes of cleavage in sphene. Intergrowths of sphene and hornblende have been observed by Groth, in the syenite of the Plauenschen Grund, near Dresden. Frequently the crystals of sphene appear cloudy or imperfectly translucent. The sections usually present very characteristic wedge-shaped forms. As a rule sections of sphene appear to be remarkably free from inclosures of other minerals.

Garnet Group.

All the members of this group crystallise in the cubic system, the common forms being either the rhombic dodecahedron or the icositetrahedron. The cleavage is parallel to the faces of the dodecahedron. The garnets vary in hardness from 6·5 to 7·5. They have a subconchoidal or uneven fracture, and they all afford an approximately white streak. Before the blowpipe most of them fuse easily, but, in accordance with the different chemical composition of the various species, they give different blowpipe-reactions with the fluxes, in which the iron reactions dominate.

In chemical composition the garnets are essentially unisilicates of different sesquioxides and protoxides. The sesquioxides are those of aluminium, iron, and chromium; sometimes also of manganese, while the protoxides are those of iron, calcium, magnesium, or manganese.

The principal sub-species have the following composition:—

1. Lime-alumina garnet, $6CaO, 3SiO_2 + 2Al_2O_3, 3SiO_2$.
2. Magnesia-alumina garnet, $6MgO, 3SiO_2 + 2Al_2O_3, 3SiO_2$.
3. Iron-alumina garnet, $6FeO, 3SiO_2 + 2Al_2O_3, 3SiO_2$.
4. Manganese-alumina garnet, $6MnO, 3SiO_2 + 2Al_2O_3, 3SiO_2$.
5. Iron-lime garnet, $6CaO, 3SiO_2 + 2Fe_2O_3, 3SiO_2$.
6. Lime-chrome garnet, $6CaO, 3SiO_2 + 2Cr_2O_3, 3SiO_2$.

The percentage of silica in the various sub-species is tolerably uniform, ranging from 35 to 40 per cent.

For descriptions of the general characters of the different sub-species of the garnet group, the student may refer to any manual of mineralogy, since only the microscopic characters of the group will be described here.

In thin sections of rocks, garnets frequently appear under the microscope merely as rounded granules, somewhat resembling spots of gum, generally colourless or of clear reddish tints, but sometimes, as in the case of picotite, of a deep brownish or reddish-brown colour, and forming irregular gummy-looking streaks. When definitely formed crystals occur, they afford, as a rule, four-sided, six-sided, and eight-sided sections. They are mostly traversed by irregular cracks. The crystals occasionally appear to have formed around a nucleus of some other mineral, such as quartz, epidote, &c. The minerals usually inclosed in garnets are magnetite, hornblende, tourmaline, quartz, occasionally apatite and augite, and colourless microliths, forming narrow prisms, whose nature has not yet been determined. . Garnets also contain at times cavities in the form of the rhombic dodecahedron (negative crystals). Although garnets occur in some eruptive rocks, yet they are most plentiful in those which have undergone strong metamorphism, and they are especially common in granulite, gneiss, talcose, and chloritic slates, and other schistose metamorphic rocks. They also occur in serpentines, and granular and crystalline limestones. The Belgian hone-stones (*coticules*) consist in great part, according to Prof. Rénard, of the manganese garnet spessartine.[1] Since they crystallise in the cubic system they exhibit single refraction. Some garnets have, however, been observed to possess double refraction, but these anomalous examples have not yet been fully investigated.

Idocrase, or Vesuvian, is in its chemical composition closely allied to the lime-alumina garnets, but crystallises in the tetragonal system. According to Sorby it often contains fluid lacunæ in which very numerous, but undetermined crystals occur.

TOPAZ

is not a mineral of common occurrence in rocks. It only attains importance as a rock-component in the 'Topazfels' of Schneckenstein in Saxony; still it is occasionally met with as an accessory constituent of certain rocks.

Its crystallisation is rhombic, but it also occurs in a granular condition. It is infusible before the blowpipe and insoluble in acids. Under the microscope it is strongly doubly-refracting; but, although it is dichroic, the dichroism is so weak that in thin sections it is scarcely perceptible. Fluid lacunæ are common in crystals of topaz; the liquid is sometimes a saline solution, sometimes liquid carbonic acid. The fluid inclosures in topaz were first investigated by Brewster in 1845, and have subsequently been examined by Sorby and Vogelsang. Rosenbusch notices the inclusion in topaz of scales of hematite, and of black flecks and granules which show no metallic lustre, and suggest the idea

---

[1] 'Mémoire sur la Structure et la composition Minéralogique du Coticule,' A. Rénard. Bruxelles, 1877, tome xli. des *Mémoires couronnés de l'Académie royale de Belgique.*

of carbonaceous matter; but, as they undergo no change when heated, the notion is regarded as erroneous.[1]

ZIRCON.

This mineral is met with in some lavas and volcanic ejected blocks, also in zircon-syenite. It crystallises in the tetragonal system. Sections of some varieties of zircon display strong dichroism, while in others it is scarcely perceptible.

In chemical composition it is essentially silicate of zirconia and since it contains no protoxides it is but little liable to undergo alteration. It, however, at times becomes hydrous, loses silica, and becomes partly replaced by other substances.

ANDALUSITE.

Crystallographic system, rhombic. The common form is a combination of the rhombic prism ∞P, the macrodomes $\bar{P} \infty$ and the basal planes oP. The crystals, especially the larger ones, are usually incrusted with and penetrated by scales of mica. At times they are feebly translucent and very rarely transparent. The cleavage is very indistinct. Andalusite sometimes occurs in a granular condition.

Its chemical composition is represented by the formula Al_2O_3, SiO_2.

Under the microscope thin sections polarise strongly and show well-marked pleochroism. Alteration is denoted by the development of a fibrous structure which usually runs in definite directions. In some crystals of andalusite carbonaceous matter occurs; but, as a rule, the mineral is very free from inclosures of foreign substances. It is most commonly met with in mica schist and slaty rocks.

The variety chiastolite or macle, so named from the spots and cruciform markings which occur in the interior of the crystals, is met with only in slates which have undergone

---

[1] Rosenbusch, *Mik. Physiog. d. Min.* p. 282. Stuttgart, 1873.

alteration from proximity to eruptive rocks, as in the Skiddaw slate where it nears the granite. The changes which take place in this district, as described by Mr. J. Clifton Ward,[1] consist first in the faint development of oblong or oval spots in the slate, together with a few crystals of chiastolite; the latter then become quite numerous and well developed, constituting the true chiastolite slate. This passes into a harder, foliated, and spotted rock which Mr. Ward regards as knotenschiefer, the spots being imperfectly developed chiastolite crystals, accompanied by more or less mica and quartz, while, in the immediate neighbourhood of the granite, the rock passes into mica-schist. Crystals of chiastolite afford sections which vary considerably in form, some giving rhomboidal outlines with a dark nucleus in the centre of the section. The boundaries of the crystal section are usually sharply defined, but the nucleus which, under the microscope, may generally be resolved into a mass of dark flakes and granules, is not very distinctly separable from the more or less translucent surrounding matter of the crystal, appearing to have a hazy boundary. Zirkel states that in such nuclei a linear arrangement of granules or flakes may sometimes be discerned passing from the centre to the angles of the section (*i.e.* to the lateral edges of the rhombic prism), but he adds that this arrangement is seldom so clearly perceptible in microscopic individuals as in the larger crystals in which such divisional markings are visible to the naked eye.

KYANITE.

Crystalline system triclinic. Chemical composition similar to that of andalusite. The crystals are generally long prisms, which appear broad in one direction and narrow

---

[1] *Memoirs of the Geological Survey of England and Wales* (The Geology of the northern part of the English Lake District), pp. 9–12. See also *Abhand. zur Geol. Specialkarte v. Elsass-Lothringen.* Die Steiger Schiefer u. ihre contactzone an den Granititen v. Barr-Andlau u. Hochwald. H. Rosenbusch, pp. 210–215. Strassburg, 1877.

in the other. The cleavages are prismatic and basal, the former being tolerably distinct parallel to the broad faces, but less so in the direction of the narrow ones. The basal cleavage enables the prisms to be easily broken in a transverse direction. The crystals are seldom terminated. It also occurs in radiating or interlacing fibrous conditions. It is transparent, with a vitreous lustre, and is coloured blue, or greyish-blue and white, individual prisms often showing a succession of blue and colourless bands which graduate into one another. The mineral is pleochroic, but this is not perceptible in thin sections. Inclosures of other minerals are rare in kyanite.

APATITE.

Crystalline system hexagonal. The crystals are usually combinations of the hexagonal prism and basis, sometimes modified by faces of the hexagonal pyramid. Although crystals of apatite several inches in length, and sometimes of much greater size, are occasionally found, still the majority of those which enter into the composition of rocks are of microscopic dimensions. These little prisms are usually very long in proportion to their breadth.

The hardness of apatite is 5. It contains from 90 to 92 per cent. phosphate of lime, and its formula is either

$$3(Ca_3P_2O_8) + CaCl_2 \text{ or } 3(Ca_3P_2O_8) + CaF_2,$$

according to whether the mineral contains chloride, or fluoride of calcium. It frequently contains both.

The detection of phosphoric acid in rocks is best effected by finely pulverising a tolerably large sample, digesting it in hydrochloric acid, filtering off the solution, and treating it with molybdate of ammonium. If phosphoric acid be present, a yellow precipitate will be formed, and the precipitation, which usually takes place very slowly, may be accelerated by frequent stirring with a glass rod. Most of the phosphoric acid which exists in rocks probably occurs in the form of apatite; except in cases where, instead of being minutely disseminated, it occurs as phosphorite.

Under the microscope apatite appears in elongated hexagonal prisms, which, when cut longitudinally, afford rectangular sections, and transversely, hexagonal ones. The boundaries of the crystals are always sharply defined. Since the mineral is uniaxial, the sections taken at right angles to the principal axis appear dark between crossed Nicols. In the colourless apatite crystals, which usually occur in rocks, no dichroism can be discerned, although in some coloured varieties of the mineral it is quite perceptible. Apatite crystals often contain light greyish or yellowish dusty matter, the nature of which has not yet been determined, although, from an examination of large crystals containing somewhat similar impurities, it has been inferred that the dust may consist partly of magnetite granules, and partly of acicular microliths, together with inclosures of glass and of fluid, the former showing motionless, and the latter movable, bubbles.

In examining some large pellucid apatite crystals from the Val Mayia, Fritzgärtner found them to contain small elongated hexagonal prisms and pores filled with liquid. The latter varied in form and size, but were mostly round. The hexagonal prisms lay with their longer axes parallel to the basal plane of the containing crystal, and appear to follow irregular curves, and to be arranged in no directions corresponding with the other axes of the crystal which contained them.[1]

Apatite crystals sometimes envelope a black opaque substance which corresponds in its boundaries with the boundaries of the containing crystal, the latter often forming little more than a clear, narrow margin around this dark nucleus. Zirkel notes the occasional symmetrical disposition of six small apatite crystals around a larger one.

[1] Private communication from Dr. R. Fritzgärtner.

Minute crystals of apatite may be distinguished from those of felspars by their hexagonal transverse sections. They may usually be distinguished from nepheline by occurring on a much smaller scale, and being of much greater length in proportion to their breadth, so that they afford rectangular sections which are generally much longer, and hexagonal sections which are much smaller, than the corresponding ones derived from nepheline. The student should also be on his guard against mistaking small apatite needles for colourless microliths of hornblende, augite, &c. Apatite crystals seem to be rather gregarious, often colonising in certain portions of a rock, and being nearly absent in others. It is of all minerals one of the most widely distributed, occurring in a vast number of rocks of very diverse mineral composition, often being present only in very minute quantity. It is even regarded by Zirkel as of more common occurrence than magnetite.

Apatite crystals usually remain clear and fresh long after the other mineral constituents of a rock have decomposed. Although so comparatively invulnerable to the natural agents which decompose rocks, it is soluble in hydro-chloric acid.

Asparagus-stone and *moroxite* are names given to yellowish green and bluish green varieties of apatite. In Canada the latter mineral occurs in a bed ten feet thick passing from North Elmsley into South Burgess. Three feet of this bed consist of pure sea-green apatite, while the remainder is made up of apatite and limestone, in which crystals of pyroxene and phlogopite also occur.[1]

Rutile

(TiO_2), which crystallises in the tetragonal system, appears deep red or brown when seen in thin section under the microscope. It is not very strongly dichroic. The crystals are often seen to be traversed by thin plates or

---

[1] *System of Mineralogy*, Dana, 5th edition, p. 533.

striæ, and by included crystals which follow the direction of the principal axis, and that of the twinning plane ∞ P. A good figure, showing this structure along the plane of geniculation, is given in Rosenbusch's 'Mikroscopische Physiographie,' vol. i. p. 187, to which work the student is referred for further particulars respecting the microscopic character of this mineral.

CASSITERITE.

(SnO_2) crystallises in the tetragonal system. The crystals, when examined in thin section under the microscope, appear by transmitted light of a honey-yellow colour.

CALCSPAR (*Calcite*) AND DOLOMITE.

Crystalline system rhombohedral. The crystals vary greatly in the combinations which they present. The most common forms are rhombohedra, scalenohedra, and hexagonal prisms, terminated by basal planes or by planes of a rhombohedron. The simple rhombohedral forms with angles of 105° 5' and 74° 55' are those which chiefly occur in rocks, and the cleavage corresponding with this form is commonly to be recognised in granular aggregates of carbonate of lime. Chemical formula $CaCO_3$.

The mineral frequently contains some magnesium or iron replacing part of the calcium. In some cases it is impregnated with sand, as in the well-known crystals from Fontainebleau, which sometimes contain over 60 per cent. of that material. When treated with acids, calcspar effervesces, giving off carbonic anhydride. (Dolomite dissolves less readily.) It is easily scratched with a knife, since it only has a hardness of 3. Before the blowpipe it is infusible, becoming strongly luminous; and, giving off its carbonic anhydride, it is reduced to quick-lime, the crystal, if previously transparent, becoming opaque, white, and pulverulent.

Under the microscope, sections of calcspar show very strong double refraction, which may be observed by using the analyser alone. The planes of cleavage which intersect one another are also generally visible, while by polarised light it is common to find that the separate granules which constitute crystalline aggregates are composed of numerous lamellæ which polarise in different colours, and which denote a system of twinning parallel to the face $-\frac{1}{2}$ R. The lines of demarcation between the lamellæ are sharply defined, and run parallel to one another in the same individual or granule ; but the planes of twinning in any one granule observe no relation to those belonging to adjacent granules. This twin structure may be well seen in crystalline limestones, statuary marble, &c. It is very characteristic of calcspar, and serves as a rule to distinguish the mineral from dolomite, which seldom shows any such structure. Reusch has demonstrated that a similar twin structure may be artificially produced in calcspar by pressure. Inclosures of other minerals are common in calcspar. The fluid inclosures which sometimes occur in calcite are generally regarded either as water, or as water containing carbonic acid or bicarbonate of lime. Dolomite may be distinguished from calcite in microscopic sections by occurring in well-formed rhombohedra, while aragonite may be known by its biaxial polarisation in convergent light.

QUARTZ.

Crystalline system hexagonal; or, as indicated by the occasionally occurring tetartohedral faces, rhombohedral. The usual forms are either hexagonal pyramids, or combinations of the hexagonal pyramid and hexagonal prism. In the former case the sections *parallel to the principal axis* yield rhomboidal figures, in the latter elongated hexagons, while in both instances the *sections transverse* to the principal axis are regular hexagons. Twinning is common, sometimes giving rise to geniculation, sometimes producing cruciform arrangements, at others causing irregular interpenetration of dissimilar parts of the crystal, the positive rhombohedral faces being irregularly penetrated by the negative, and *vice versâ*. The chemical composition of quartz is SiO_2, with occasional impurities, such as iron oxides, titanic acid, &c. Quartz is infusible before the blowpipe, insoluble in all acids except fluoric acid. It is also more or less acted upon by a hot solution of potash; in the case of the purer crystallised varieties, but very slightly. In the compact and crypto-crystalline conditions its solubility in this reagent is, however, according to Rammelsberg somewhat greater.[1] The hardness of quartz is 7, the point of a penknife producing no effect upon it, unless it be in a finely granular condition, when the point may simply rake up and detach a few granules, upon which, however, it is unable to make any impression. In this instance, as in others, it behoves the student to be on his guard in testing the hardness of granular or finely crystalline substances, to distinguish between the disintegration of granular structures and true streak. The fracture of quartz is conchoidal. The specific gravity $= 2\cdot 65$.

Sections of quartz appear clear and pellucid under the microscope, and polarise in strong colours. Moderately thick sections cut at right angles to the principal axis, show circular polarisation when examined in convergent polarised light. The plane of polarisation is sometimes right-handed, sometimes left-handed, in its rotation, and both the right-handed and left-handed phenomena are at times seen in the same crystal. In right-handed crystals the rings appear to expand, and in the left-handed to contract, when the analyser is turned to the right.

Quartz is seen frequently to contain inclosures of other substances, sometimes as crystals, sometimes in the form of lacunæ filled with liquids, &c. These inclosures are often visible to the naked eye, but the microscope commonly reveals their presence in vast numbers. The crystals of

[1] *Pogg. Ann.* cxii. 177.

most frequent occurrence are those of rutile and chlorite; crystals of kyanite are also occasionally met with. Lacunæ of glass, others filled with gas, and others containing portions of the rock matrix in which the crystals are imbedded, are by no means uncommon in some rocks. The most numerous lacunæ, however, are those containing liquids The liquids are frequently pure water; sometimes water holding carbonic acid in solution, sometimes liquid carbonic acid, sometimes a supersaturated solution of chloride of sodium, minute crystals of rock salt being visible within the lacunæ under tolerably high powers. These lacunæ are at times completely filled with the fluid, at others they are seen to contain bubbles which vary in magnitude and are regarded as representing the diminution of volume which the fluid in the cavity has undergone during the cooling of the rock mass in which it occurs. Deductions based upon the relative volumes of the fluids, and of the vacuities in such cavities, may be found in the paper communicated to the Geological Society by Mr. Sorby in 1858. It is, however, deserving of note, as pointed out by Mr. John Arthur Phillips, that, in the same crystal, cavities may be found, some completely filled, while others contain vacuities whose relative volumes to those of their surrounding fluids vary very considerably. The bubbles in these lacunæ are often moveable, being displaced either by simply turning the crystal, or more usually by heating it, in which case the bubble undergoes diminution of volume, or even disappears. In some small lacunæ diminutive bubbles, which have a spontaneous motion, are visible under high magnifying powers. These bubbles are, however, so minute that they appear frequently as mere specks, and it needs careful and steady watching to see their motion, which looks like a tremulous gyration, often resulting in a comparatively well-marked change of place, followed perhaps by a pause, to be again succeeded by the oscillations and gyrations already mentioned.

Tridymite is a form of silica discovered in 1866 by Vom Rath. It occurs in very small six-sided tabular crystals. The system to which these crystals belong has not yet been satisfactorily determined. The specific gravity is 2·2 to 2·3, the same as that of opal. The crystals occur in compound groups, mostly composed of three individuals, whence the name. It is found in the sanidine oligoclase trachyte of the Drachenfels, in a volcanic porphyry from near Pachucha in Mexico, in a porphyry from Waldböckelheim, in some Hungarian liparites, in the hornblende-andesites of Dubnik,[1] in a trachytic-looking phonolite from Aussig,[2] in the Wolf Rock (phonolite) described by Allport, in several phonolites described by Möhl.[3] It has also been mentioned by A. v. Lasaulx as occurring in the trachyte of Tardree, in Ireland. According to Vom Rath, tridymite is optically uniaxial, but V. Lasaulx regards it as biaxial and triclinic. It may be identical with the Asmanite of Maskelyne.

A globular condition of silica has been lately described by Michel Lévy[4] as occurring in the euritic porphyries of Les Settons, and similar globular conditions of silica have also been observed and noticed by M. Vélain in a quartz-trachyte from Aden. The former author regards this condition as intermediate between the crystallised and the colloid forms of silica.

The following extract from M. Michel Lévy's paper will convey an idea of the microscopic characters of these globules: 'Between crossed Nicols one is surprised to see such regular globules, the centre of each appearing to be a pole of symmetry with four extinctions situated at right angles for every total revolution of the section; one is therefore forced to conclude that they are composed of a crystallised substance, and are, moreover, orientated in an unique manner.

---

[1] H. Rosenbusch, *Mikroskop. Physiogr. d. massigen Gesteine*. Stuttgart, 1877, p. 301.
[2] *Ibid.* p. 225.
[3] H. Möhl, *Basalte u. Phonolithe Sachsens*. Dresden, 1873.
[4] *Bull. Soc. Géol. de France*, 3ᵉ série, t. v. 1877, no. 3.

Sometimes the extinction is simultaneous over an entire globule, sometimes it is different for two or more segments; but the most curious peculiarity, exhibited by the concentrically-zoned globules, lies in the fact that two adjacent zones do not always undergo extinction simultaneously. Such a globule will undergo extinction in its central portion and will, at the same time, present a perfectly regular narrow border which is still illuminated; if then the observer continue to turn the section, this border will become dark while the spherical central nucleus will in its turn become clear in a homogeneous manner.'

MAGNETITE.

Crystalline system cubic. It usually occurs in the form of the octahedron, sometimes in that of the rhombic-dodecahedron, also granular and massive. Cleavage parallel to the faces of the octahedron. Colour black. Streak black. Strongly magnetic and often displays polarity. The chemical formula of magnetite is FeO, Fe_2O_3, or Fe_3O_4. The approximate percentage composition is $Fe_2O_3 = 69$. $FeO = 31$. Magnetite is frequently titaniferous. It is very difficultly fusible before the blowpipe. When pulverised it is completely soluble in hydrochloric acid. Even in the thinnest sections magnetite appears opaque under the microscope; nevertheless, when it has undergone alteration either into hematite or limonite it appears feebly translucent at times and of a reddish or brownish colour. The sections of magnetite crystals, which are of most common occurrence in rocks, present square forms which represent sections passing through opposite solid angles of the octahedron, or triangular forms which result from sections taken parallel to one of the faces of the octahedron (fig. 69). Twin crystals sometimes occur, the twinning taking place on a plane parallel to a face of the octahedron. The superposition of one crystal on another sometimes gives rise to cruciform figures. Magnetite

FIG. 69.

very usually occurs in a granular condition in rocks, sometimes in coarse irregular grains, sometimes as a fine dust, while at others these granules form segregations which give rise to rod-like forms. Occasionally, as in some basalts, magnetite crystals are grouped in a very regular manner, following lines which are frequently disposed at right angles. Such groupings are not merely to be met with in volcanic rocks, but also in furnace slags, and their arrangement often seems to imply the rudimentary stages of aggregation which might eventually result in the formation of a large crystal from the contiguous development of smaller ones.

TITANIFEROUS IRON.

Crystallisation rhombohedral. It mostly occurs in tabular forms with the basal planes largely developed and with hexagonal boundaries. Titaniferous iron is opaque and black, with a semi-metallic lustre.

Chemically it has been regarded by G. Rose as an isomorphous mixture of titanium and peroxide of iron. Rammelsberg, however, gives it the general formula of $(mFeTiO_3 + nFe_2O_3)$—magnesium usually replacing some of the iron: one specimen has yielded as much as 15 per cent. of magnesia. When heated alone before the blowpipe titaniferous iron is infusible. Heated in concentrated sulphuric acid it affords a blue colouration but is insoluble. When pulverised it is, however, soluble in nitro-hydrochloric acid.

When occurring in microscopic preparations it is often difficult to distinguish it from magnetite, except when it affords well-marked rhombohedral sections (fig. 70). It may, however, often be recognised by the peculiar greyish-white alteration-product which is often developed within it, and which frequently follows definite crystallographic directions. When this alteration is far advanced, merely a dark skeleton or a few dark specks of the unaltered mineral remain, surrounded by

FIG. 70.

the white decomposition product. The precise nature of this white substance has not yet been ascertained, but it is generally assumed to be either titanic acid or some silicate of titanium.[1]

Both titaniferous iron and magnetite frequently occur together in the same rock.

HEMATITE.

The crystallised variety, specular iron or ironglance, belongs to the rhombohedral system, and mostly occurs in six-sided, thin, tabular crystals in which the basal planes are largely developed, while the boundaries are formed either by faces of the rhombohedra R and $-\frac{1}{2}R$, or by faces of a hexagonal prism. Crystals of this kind may be easily procured by dissolving a fragment of the mineral carnallite, when the residue will be found mainly to consist of beautifully-developed thin tabular crystals of specular iron, which are translucent, and of a clear red or orange-red colour. Thicker crystals appear black or iron-grey, and, as in some of the specimens from Elba, show beautiful superficial iridescence. Sometimes the crystals are only imperfectly developed, or merely form irregularly-shaped scales. In this scaly condition it is spoken of as iron-mica or micaceous hematite (Eisenglimmer). Hematite also occurs in a granular state, sometimes earthy as reddle, while reddish stains of ferric oxide are of common occurrence in rocks; especially in those which have undergone weathering.

The botryoidal or mammillated forms of hematite mostly occur in pockets or cavities in rocks into which they have been subsequently introduced, or else line drusy cavities, but hematite in this form does not occur as a common rock constituent, although in a compact and massive condition it is often met with in lodes. In some cases, however, the massive and micaceous forms of hematite may almost of

[1] It has since been examined and described by Gümbel, under the name of leucoxene.

themselves be regarded as rock masses; a hill in the state of Missouri (the Pilot Knob, 700 feet high) consisting almost exclusively of hematite.

Hematite gives a blood-red or cherry-red streak. It is feebly magnetic. Its chemical composition is Fe_2O_3 when pure, but it is often rendered impure by admixtures of sand, clay, &c. Before the blowpipe it is infusible, but becomes black and strongly magnetic when heated in the reducing flame.

Under the microscope it is usually seen to occur in irregular flecks and scales, distinct crystalline forms not being of common occurrence in rocks.

It exhibits no dichroism, and shows red tints of various intensity by transmitted light. By reflected light it also usually appears red, especially when in an earthy or finely granular condition.

LIMONITE.

This is a hydrated peroxide of iron having the formula $2Fe_2O_3, 3H_2O$. It is essentially a decomposition product, resulting from the alteration of protoxides, or of anhydrous peroxides of iron, which have previously existed as constituents of other minerals, or in the latter case sometimes simply as hematite itself. Limonite occurs in stalactitic, mammillated, pisolitic, or earthy, conditions. It is commonly blackish-brown or yellowish-brown, in an earthy or ochreous state often yellow. The streak is yellowish-brown. In thin sections of rocks it is often seen to occur, forming pseudomorphs after crystals of various ferruginous silicates, and as irregularly-shaped blotches. It appears opaque under the microscope, or occasionally, in very thin sections, it is feebly translucent, and of a brownish colour.

IRON PYRITES.

Crystallises in the cubic system, the most common form being the cube. The faces of the crystals are frequently striated, the striæ on one face lying at right angles to those on the adjacent faces. Pyrites also occurs massive, in nodules which have internally a radiating structure, (many of these may no doubt be referred to marcasite), while in some rocks it exists in a granular or finely-disseminated state, sometimes forming pseudomorphs after other minerals. Fossils are at times entirely replaced by pyrites. It is mostly of a pale brass-yellow colour, gives a greenish or brownish-black streak and a conchoidal or uneven fracture. It has a strong metallic lustre, strikes fire with steel, and fuses before the blowpipe to a metallic globule which is attractable by the magnet. When heated it gives off sulphur. When fused with carbonate of soda, the assay, if placed on a clean silver surface, and moistened with a drop of water, produces a dark stain on the silver. Its chemical composition is iron $= 46\cdot 7$, sulphur $= 53\cdot 3$, giving the formula FeS_2.

Under the microscope, in thin sections of rocks, pyrites appears perfectly opaque. The ground surfaces look glistening and yellowish by reflected light, and this partly serves to distinguish it from magnetite.

The sections are those resulting from cubes or dodecahedra sliced in various directions, except in cases where the mineral is pseudomorphous after some other mineral. Occasionally pyrites occurs in minute elongated rod-like forms.

Marcasite resembles pyrites, except that it crystallises in the rhombic system. Twinning is common in this species.

COPPER PYRITES (CHALCOPYRITE).

This mineral is occasionally met with in rocks such as diabase, some granites, gneiss, argillaceous schists, &c. It crystallises in the tetragonal system; the crystals, however, closely approximating to cubic forms. It usually has a deeper yellow colour than iron pyrites, from which it may be distinguished by its inferior hardness, being sectile, while iron pyrites cannot be cut with a knife. Copper pyrites does not emit sparks when struck with steel. Before the blowpipe it colours the borax bead blue in the oxidising flame, but to get this colouration the assay should not be previously reduced, for, if so, only a deep green colouration will be procured. The blue colour is probably due to sulphate of copper, and a previous roasting of the assay of course expels the sulphur. It is soluble in nitric acid, with the exception of the contained sulphur, forming a green solution which changes to a deep blue on the addition of ammonia in excess.

The chemical composition of copper pyrites is copper $= 32\cdot 5 - 34$. Iron $= 29\cdot 75 - 31\cdot 25$. Sulphur $= 34 - 36$ per cent. The formula is $Cu_2 S, Fe_2 S_3$.

Under the microscope it appears opaque. By reflected light it shows a somewhat metallic lustre on ground surfaces, and is generally rather deeper in colour than iron pyrites, but not much reliance can be placed upon this appearance, and its presence should be confirmed by chemical examination.

ZEOLITES.

Want of space precludes more than a brief mention of the microscopic characters of a few of the most common zeolites. They may all of them be regarded as alteration products, and in all probability never form normal constituents of rocks. They usually occur either lining or completely filling cavities in vesicular and other volcanic rocks, and also occupy fissures and small cracks; occasionally they are developed in crystals of other minerals which have undergone more or less alteration. They often occur in spherical crystalline aggregates, with a radiating structure, in which case they exhibit a black cross when examined between crossed Nicols, the arms of the cross coinciding with the planes of vibration of the Nicols. The section may be horizontally rotated while the crossed Nicols remain stationary, yet, although the object revolves, the dark cross does not move. This is explained by Groth as being due to the principal directions of vibration of the doubly-refracting fibres lying parallel and at right angles to their longer axes, and bearing a similar relation to rays which undergo extinction between the crossed Nicols. If the analyser be turned through 10° or 20° the dark cross becomes somewhat faint, and a second imperfectly developed cross appears, which makes an angle of 5° or 10° with the fixed cross. It will therefore be seen that it travels at only half the rate of rotation of the analyser. When the analyser has been so far turned that the two Nicols stand parallel, the first cross disappears and the second imperfect cross attains its maximum intensity. This phenomenon is met with in all doubly-refracting, radiate crystalline aggregates; and, since zeolites frequently occur in this condition, its presence in certain rocks often suggests that such aggregates are zeolitic.

Natrolite, which crystallises in the rhombic system, possesses weak double refraction, and polarises in vivid colours. It very commonly occurs in crystalline aggregates, which almost invariably have a radiate structure, and then show, especially when in rounded masses, the interference figure characteristic of such aggregates. At times, also, natrolite is seen filling minute fissures. In this case crystallisation commences on either side of the fissure, and the crystals meet in the middle, their termination giving rise to a zig-zag median line which divides the two growths. Nepheline crystals at times become partly altered into natrolite, a meshwork of little prisms of natrolite, in some instances, almost completely filling the crystal.[1]

Analcime, so far as is yet known, crystallises in the cubic system, but, although regarded as cubic, it exhibits some rather exceptional optical properties, first pointed out by Brewster, and subsequently investigated by Descloizeaux,

[1] Rosenbusch, *Mik. Physiog.* (Min.) vol. i. p. 285. Stuttgart, 1873.

Rosenbusch, and other observers. According to Descloizeaux, sections cut parallel to any one of the faces of the cube, when viewed in the direction of one of the axes by parallel polarised light, appears between crossed Nicols perfectly dark in the direction of the two other axes, while, in the direction of the diagonals of the cube-face, a faint bluish, distorted cross appears. Analcime is seldom or never a normal constituent of rocks. Tschermak, however, regards it as an essential component of the rock which he terms teschenit, which consists of plagioclase, hornblende, analcime, magnetite, biotite, and apatite. In a leucitophyr from Rothweil, near the Kaiserstuhl, analcime occurs pseudomorphous after leucite; at all events, the leucite crystals contain fibrous and granular aggregates of analcime, which at times almost totally replace them.

Heulandite has been observed to contain various microscopic inclosures such as minute orange-yellow coloured acicular crystals, irregularly-shaped or round granules and flecks of a reddish-yellow mineral, and, in one specimen, from the Faroe Isles, Rosenbusch noted the occurrence of innumerable perfectly-developed microscopic quartz crystals. The colour of heulandite is due to the reddish and orange-yellow inclosures just alluded to. They have been regarded as göthite, limonite, or hematite. Zirkel considers that they are hematite.

Chabasite.—This mineral appears from numerous observations always to be devoid of fluid inclosures. Microscopic envelopments of quartz have been met with in chabasite.

For further particulars respecting the large family of zeolites the student is referred to the various manuals and text-books of mineralogy.

CRYSTALLITES.

Under this head may be grouped a vast number of purely microscopic bodies, which, in their progressive development, represent the various forms and conditions of mineral matter, from its departure from an amorphous state, to one of crystallographic completeness, such as may be correlated, if not identified, with the crystals of recognised mineral species. The forms belonging to the highest stage of this microscopic development are spoken of as microliths, and they frequently present crystal faces sufficiently distinct to admit of goniometric measurements, and optical characters well enough defined to permit their correlation with recognised minerals. The less perfectly developed crystallites cannot however be referred to any particular species, and hence has arisen the necessity for the employment of various terms, more or less vague, and each of them embracing a vast multitude of different forms, but convenient, because indicative of structural types. Doubtless, as knowledge increases, these terms will give place to better ones with more precise significations, and the progressive developments which these minute forms display will be properly worked out, and afford a key to the important subject of crystallogenesis.

The crystallites may be ranged in a descending series as follows:

>Microliths.
>Crystalloids.
>Trichites.
>Globulites.

The *globulites* represent the most embryonic stage of crystallogenesis, the most rudimentary change effected in amorphous matter. They are spherical in form, and by their coalescence give rise to variously shaped groups, according to the number of individual globulites which enter into their composition. Sometimes they arrange themselves in strings, and into other systems of disposition, implying more or less symmetrical grouping. They usually show a central speck or nucleus, and at times display concentric markings and indications of a radiate structure.

Trichites (from θρίξ, a hair) are minute elongated bodies resembling small hairs or fibres; sometimes they are straight, sometimes they cross one another in a more or less regular manner; at others they appear bent in zigzags, or are curiously twisted, while occasionally a number of trichites emanate from a central granule around which they radiate or twirl like whip-lashes. Some trichites show regular or interrupted lines of granules attached to them, forming rows like beads either upon one or upon both sides of the trichite.

The crystallites proper and crystalloids exhibit in many instances a much higher development, being bounded by curved or by straight lines, and often assuming crucial or stellate forms, which appear to result from the symmetrical grouping of individual crystallites. The crystalloids especially exhibit considerable complexity of internal structure, while in external form they often approximate to crystals of recognised minerals. Some of them indeed show so close a resemblance to true crystals that one cannot help feeling impressed with the significance of their internal structure when contrasted with that of larger crystals.

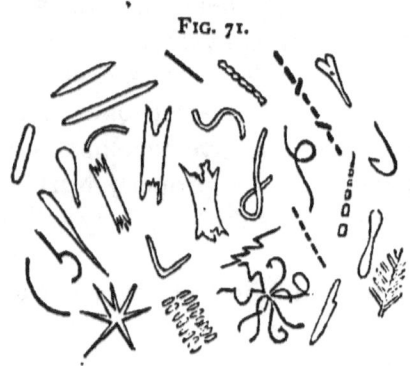

FIG. 71.

The accompanying figures convey a far better idea than any description could of the forms which these minute bodies present.

Microliths.—These again show a more complete phase of development than the preceding forms. They are sometimes very imperfectly developed, but in all cases it is generally considered that they exhibit a nearer approximation to true crystals. They very commonly display double refraction, occasionally show hemitropy, and frequently present sufficiently well-developed faces to enable the observer to measure their relative inclination. In some of the larger microliths dichroism may now and then be detected. It is therefore possible at times to determine with some precision the species to which a microlith belongs. Occasionally crystals are to be met with which are visibly built up of microliths, as in the case of the hornblende crystal, fig. 72, which is copied from a woodcut in Zirkel's 'Mikroskopische Beschaffenheit der Mineralien und Gesteine.' Microliths are to be found in most eruptive rocks, and in many metamorphosed sedimentary deposits. Globulites, trichites, crystallites, and crystalloids may best be studied in sections of vitreous rocks such as obsidians, pitchstones, and perlites, also in artificially formed glasses and slags. Streams of microliths may commonly be seen under the microscope in sections of pitchstone and perlite. They often lie with their longest axes in one direction, which represents the direction of flow in the once viscid mass, for we not merely see microliths but also strings of vitreous matter, spherulites, &c., elongated and drawn out in the same direction. The microliths sweep in curves round any large crystals or fragments which may chance to lie in their course, and seem to have behaved just as planks or sticks do when floating down a stream. These appearances in a rock are spoken of as fluxion structure or fluidal structure.

FIG. 72.

The development of microliths is one of the causes of devitrification in glassy rocks and in artificial glass. Microliths also occur as products of alteration, frequently filling or partially filling the interior of crystals which are undergoing decomposition.

For further information upon the microscopic characters of crystallites, both of natural occurrence and of artificial formation, the reader is referred to 'Die Krystalliten' by the late Hermann Vogelsang. Bonn, 1875.

Fluid Inclosures, &c.

When salts are allowed to crystallise from a saturated solution, it is by no means uncommon to find that the crystals, in the course of their formation, shut in small portions of the mother-liquor; and should the temperature at which the crystals form be a moderately high one, the imprisoned fluid will, upon cooling, diminish in volume, so that a vacuity in the form of a bubble will also be seen to occupy a portion of the cavity originally filled by the liquid. The relative dimensions of these bubbles to the cavities which contain them have been carefully studied by Sorby, Rénard, Phillips, Ward, and other observers. The cavities which contain these fluids are of very variable form (fig. 73). Occasionally they are so large as to be distinctly visible to the naked eye, but usually they are of quite microscopic dimensions. In the exceptionally large ones the bubble may be seen to move to different parts of the cavity by merely turning the crystal about in the hand. In the microscopic cavities the bubble can be made to move and the liquid to expand by the application of heat. This may be effected either by means of a voltaic current or by a blast of heated air.[1] The cavities containing air and gases, which are sometimes met with in crystals, present strong, dark outlines, which serve to distinguish them from those containing fluids, while the differences in the refractive indexes of their contents also serve as another distinction. Furthermore, although some cavities occur completely filled with liquid,

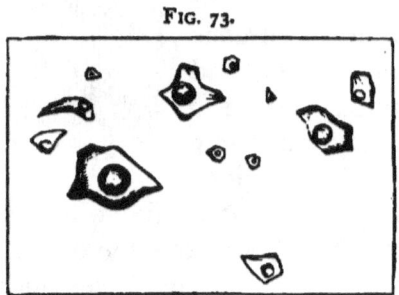

FIG. 73.

---

[1] W. N. Hartley, 'On Identification of Liquid Carbonic Acid in Mineral Cavities,' *Trans. Royal Mic. Soc.*, vol. xv. p. 173, 1876.

still the presence of movable bubbles in most of the fluid-containing cavities at once affords a means of distinguishing them. In some microscopic inclosures of fluid very minute bubbles, which have a spontaneous motion, may be seen under high powers.

The liquids usually contained in such cavities are water, liquid carbonic acid, and aqueous solutions of salts, frequently of chloride of sodium; and occasionally cavities may be seen in quartz which, besides the liquids and bubbles, contain minute cubic crystals of rock salt (fig. 74).[1]

FIG. 74.

Glass Inclosures are of common occurrence in the minerals which are met with in vitreous rocks, or in rocks which contain a certain amount of interstitial glassy matter. They are spherical, spheroidal, fusiform, or of very irregular shape, or else they assume definite crystallographic forms, corresponding as a rule with that of the crystal in which they occur. Such forms may be regarded as negative crystals. They either appear as singly refracting matter, or, when more or less devitrified, as doubly refracting. In the latter case they may be devitrified either by the development of crystalline granules or of microliths. Glass inclosures frequently contain bubbles, but these bubbles are fixed, and do not change their position when the section is heated.

Stone Inclosures are analogous to the foregoing, except that they consist of portions of a rock's magma which has a crystalline and not a vitreous character.

In the cases both of glass and stone inclosures small portions of the matrix have been taken up while still in a fluid or pasty condition by the crystals in which they occur,

---

[1] Vide *Memoire sur les Roches dites Plutoniennes de la Belgique*, De la Vallée Poussin et A. Rénard. Bruxelles, 1876. Also 'The Eruptive Rocks of Brent Tor and its Neighbourhood,' *Memoirs of the Geological Survey of England and Wales*, F. Rutley. London, 1878.

and the crystal having developed itself before the solidification of the surrounding magma, these small portions have been shut off and imprisoned. Sometimes the severance of the little mass of glass or other matrix has not been perfectly effected, and it merely appears as a small pocket with a constricted neck, which opens out on the margin of the crystal. Such partial inclosures may frequently be seen in the quartz of quartz-porphyries and quartz-trachytes.

Provisional Names applied to Minerals.

The following are terms used to designate provisionally certain substances which are sometimes met with in thin microscopic sections of rocks, and which, from occurring only in very minute quantities difficult of isolation, it has not as yet been possible to analyse. Their precise chemical constitution and mineralogical affinities are, therefore, undetermined; and, to avoid erroneous descriptions of them, certain terms have been coined which merely imply substances which present certain microscopical appearances, and whose mineralogical characters may vary more or less, and may embrace several distinct mineral-species under each term.

Viridite includes mineral matter which is probably referable to different varieties of chlorite and serpentine. It appears under the microscope in the form of translucent green matter, either in little scales, or fibrous aggregates. It may always be regarded as a product of decomposition, and frequently represents the alteration of such minerals as hornblende, augite, olivine, &c. It is probably a silicate of magnesia and protoxide of iron.

Opacite is the term applied to perfectly opaque, black, amorphous, microscopic granules, patches, or scales. It is usually present in rocks which contain magnetite; frequently it forms pseudomorphs after other minerals. It is regarded by Zirkel as representing earthy silicates, possibly allied to micas, and amorphous metallic oxides, especially hydroxides and oxides of titanium and manganese. In some cases it may be amorphous magnetite, or carbonaceous matter, graphite, &c.

Ferrite is amorphous red, brown, or yellow earthy matter which is often pseudomorphous after ferruginous minerals. Chemically it most likely represents hydrous or anhydrous oxides of iron, but the different kinds cannot be referred with any certainty to definite mineral species.

FELSITIC MATTER.

This substance, which is of such common occurrence in many rocks, and, in some, constitutes a very large proportion, forming the groundmass of quartz-porphyries and many other porphyritic rocks, and often representing, in the rhyolitic series, the devitrification of glassy matter, has hitherto been described in a more or less vague manner by numerous observers. The student has consequently been left in a state of considerable doubt as to what felsitic matter really is, and, as a rule, the more he has read on the subject, the less able has he been to fix any satisfactory definition for the term.

A masterly account of the various opinions which have been put forward on this subject will be found in Rosenbusch's 'Mikroskopische Physiographie der massigen Gesteine.' Stuttgart, 1877, p. 60 *et seq.* In this place it will suffice to give the conclusions arrived at by Prof. Rosenbusch, since, although they represent in part the views of Prof. Zirkel, they seem to meet most of the objections to which other definitions are open, and possess a precision hitherto wanting in most descriptions of these difficultly determinable substances.

To begin with ; these substances which cannot be properly investigated, except under high magnifying powers, may be resolved microscopically either into a thoroughly crystalline aggregate, or into homogeneous, amorphous matter.

The former is designated *groundmass* by Vogelsang and Rosenbusch; and the latter *magma*.

Zirkel employs the name *basis* for the amorphous substance, or 'unindividualised ground-paste,' as he terms it, and Rosenbusch also adopts the term basis.

Zirkel's definition of a micro-felsitic basis is: 'that it is amorphous, that it shows, in thin sections, no independent contours. Its boundaries are moulded upon the forms of the crystalline constituents, and it forms roundish creeks or inlets in the latter. Its true nature is variable, and not easy to render in words. It represents a devitrification product in which, indeed, a hyaline aspect is utterly wanting, but which, on the other hand, is not separable into true individualised parts. It generally consists of quite indistinct, often half-fluxed granules, or indistinct fibres which constitute the micro-felsitic mass. Between crossed Nicols it becomes, in its typical development, perfectly dark, but also, at times, transmits a very feeble, fluctuating light. The little fibres and granules often show a decided or a rough tendency to radial arrangement. In thin section micro-felsitic matter appears very clear, and either light-greyish, yellowish, reddish, or quite colourless. It is often speckled with little dark granules which in certain spots show a crude radial arrangement, or it may be sprinkled with brownish-yellow and brownish-red granules of a ferruginous mineral.

An ultimate glass magma may be present in many micro-felsitic masses, although not clearly to be recognised as such. Experience shows that weathered felspars may be represented by micro-felsitic matter.[1]

Rosenbusch states that in many cases felsite, or the groundmass of porphyries, consists of a microscopically fine-grained aggregate, formed of minerals which can be identified with those constituting granitic rocks, often in the same combinations as those in which they occur in rocks

---

[1] *Mikroskop. Beschaff. d. Min. in Gest.*, Zirkel. Leipzig, 1873, p. 280.

of the granitic group. One or other of these minerals is often absent, and of these mica is the one which is generally missing. So long as the granules of such aggregates, which differ in no essential respect from many vein-granites, or the groundmasses of many granite-porphyries, are not of too minute dimensions, one can recognise the mosaic-like aggregate polarisation and the sharp boundaries of the individual crystalline grains. Diminution in the size of the grains naturally renders the recognition of the individual particles more difficult, and often impossible. The individual granules do not merely lie side by side, but also in various planes one over another, and the various refractions, reflections, and interferences which ensue from these overlaps tend to render any deductions concerning the optical characters of the constituent granules highly untrustworthy, and give rise to the generally vague transmission of light which characterises these aggregates when they are examined between crossed Nicols.

Rosenbusch goes on to state that if we accept Groth's definition of a crystal as a compact body in which the elasticities differ in different directions, and, if we furthermore allow that external boundaries are immaterial so far as the foregoing definition extends, it follows that if double refraction be the result of any mechanical conditions of tension, or strain, the expression 'non-individualised,' used by some authors in reference to micro-felsitic matter, is either meaningless, or that it indicates, at best, that external form does not entail special internal conditions.

From this point of view Rosenbusch designates as *crystalline* all those parts of felsites which are doubly refracting, so long as it cannot be demonstrated that their anisotropy is in any way the result of anything resembling conditions of strain which are not related to molecular structure.

Basing his nomenclature upon these considerations, Rosenbusch describes those parts of porphyritic groundmasses which are aggregates of mineralogically-recognisable elements as *micro-crystalline*, while those parts which are simply crystalline aggregates, without any definite character being discernible in the constituent particles, he terms *crypto-crystalline*.

Those portions of porphyritic groundmasses in which no double refraction can be recognised must be regarded as amorphous, although, as Rosenbusch remarks, that, excluding isometric crystals, there are those belonging to other systems in which the elasticity-differences in some directions become too insignificant to afford any perceptible phenomena of double refraction when thin sections of them are examined. He also cites certain alteration-products after pyroxene and amphibole, in which their anisotropic character can only be distinguished by their pleochroism. In very many cases micro- or crypto-crystalline matter contains an intimate admixture of fine films, stripes, and flecks of a perfectly structureless and almost invariably colourless substance which remains dark in all positions between crossed Nicols. It may be absolutely homogeneous, or it may contain excessively minute granules and trichitic bodies of various kinds. This substance Rosenbusch designates *glass* or *glassy-basis*. The condition in which it contains the granules and trichites he regards as a phase of devitrification. In most instances this impure or devitrified matter is opaque, or so feebly translucent, and occurs in such minute films or grains, that a determination of its isotropic or anisotropic character is seldom possible.

Instead of a true glassy-basis, matter of a somewhat different kind is very often present, forming excessively thin films which appear interwoven with the micro- or crypto-crystalline aggregates. This substance is perfectly isotropic, colourless, greyish, yellowish, or brownish, but, unlike true glass, it is not structureless, but appears to be made up of extremely minute scales, fibres, granules, or aggregates of granules, together with other developed forms and interstitial matter. It differs also from micro- and crypto-crystalline aggregates in its want of any action upon polarised light. This substance is the *micro-felsite* or *micro-felsitic basis* of Rosenbusch. It is not micro- or crypto-crystalline, and it is not amorphous in the sense in which those terms are employed.

Rosenbusch adds that it yet remains to be shown whether micro-felsitic matter is inert upon polarised light, owing to exceptional conditions of tension, or whether it should be regarded as a fibrous, scaly, or granular glass, or as something else.

The observations of Leopold von Buch, Delesse, Stelzner, Wolff, Vogelsang, Allport, Kalkowsky, and other writers, are commented upon in the review of this subject, which Rosenbusch gives in the work from which these statements have been extracted.[1]

TABLE SHOWING THE CLEAVAGES OF THE MOST COMMON ROCK-FORMING MINERALS.

Cubic System.

	Dodecahedron ∞O
Garnet	very imperfect
Hauyne	rather perfect
Nosean	,,
Sodalite	,,

Tetragonal System.

	Prism (2nd order) $\infty P \infty$	Prism (1st order) ∞P	Basis oP	Pyramid (1st ord.) P
Leucite	most imperf.	—	most imperf.	—
Scapolite	rather perfect	less distinct	—	—
Zircon	—	imperfect	—	imperfect
Melilite	—	—	rather perfect	—

---

[1] *Mikroskopische Physiographie.* Stuttgart, 1877, bd. ii. p. 60 *et seq.*

Hexagonal System.

	Rhombohedron R		Prism ∞ P	Basis 0 P
Biotite	—	—	—	highly perf.
Apatite	—	—	imperfect	imperfect
Tourmaline	very imperf.	—	∞P2 very imperf.	—
Calcspar	very perfect	—	—	—
Nepheline	—	—	imperfect	imperfect
Quartz[1]	most imperf.	—	traces have been observed	—

Rhombic System.

	Brachy-pinakoid ∞ P̆ ∞	Macro-pinakoid ∞ P̄ ∞	Prism ∞ P	Basis 0 P
Olivine	rather distinct	very imperf.	—	—
Enstatite	imperfect	—	distinct	—
Bronzite	very perfect	traces	imperfect	—
Hypersthene	,,	very imperf.	distinct	—
Andalusite	traces	traces	not very dist.	—
Muscovite	—	—	imperfect	highly perfect

Monoclinic System.

	Ortho-pinakoid ∞ R̶ ∞	Clino-pinakoid ∞ R̶ ∞	Prism ∞ P	Basis 0 P
Orthoclase	—	very perfect	very imperf.	very perfect
Augite	imperfect	imperfect	more or less perf.	—
Diallage	perfect	,,	rather perf.	—
Hornblende	very imperf.	very imperf.	very perfect	—
Epidote	perfect	—	—	very perfect

Triclinic System.

	Brachy-pinakoid ∞ P̆ ∞	Macro-pinakoid ∞ P̄ ∞	Prism ∞ P	Basis 0 P
Labradorite	rather perfect	—	—	very perfect
Oligoclase	,,	—	hemipr. imperf., also pr. imperf.	perfect
Anorthite	perfect	—	—	,,
Albite	,,	—	hemipr. imperf.	,,[2]

---

[1] Cleavage in quartz is rare.   [2] Also tetarto-pyram. imperfect.

PLATE II.

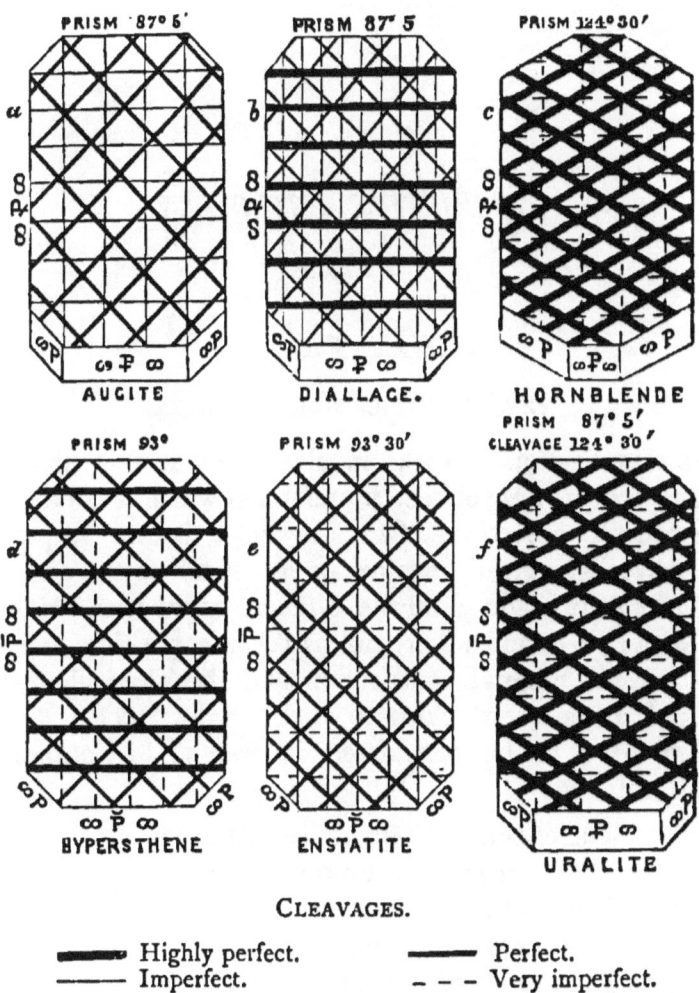

CLEAVAGES.

▬▬ Highly perfect.   ▬ Perfect.
── Imperfect.   - - - Very imperfect.

The observer is supposed to be looking down directly on the basal planes of the crystals.

Except in fig. *f*, the angles of the prismatic cleavage correspond with those of the prisms.

Figs. *a*, *b*, *c*, and *f* represent monoclinic crystals. Figs. *d* and *e* represent rhombic crystals.

PART II.

DESCRIPTIVE PETROLOGY.

CHAPTER XI.

THE CLASSIFICATION OF ROCKS.

THE classification of rocks involves considerable difficulty, and no scheme has yet been propounded which is not more or less open to objection. Our knowledge is not at present extensive enough to enable us to speak with certainty regarding the origin of all the different rocks with which we are acquainted, and we are not as yet in a position to assert how far the mineral constitution and the physical characters of rocks afford clues to their origin. The following points have to be considered in framing a petrological classification :—

(i.) The chemical composition of the rocks.
(ii.) Their mineral constitution.
(iii.) Their physical characters.
(iv.) Their mode of occurrence.
(v.) Their order of sequence in time.

The chemical examination of a rock shows us what elementary substances enter into its composition, and may afford some clue to its mineral constitution and to its origin.

The mineralogical and physical examinations teach us how those elementary substances have combined, and, in some cases, the conditions under which those combinations have been effected, the various minerals which enter into the composition of the rock, the crystallographic and other physical peculiarities which the component minerals present, the relative order in which those minerals have sometimes crystallised, the arrangement, if any, which the individual crystals, granules, scales, or fragments observe towards each other, and the general state of aggregation of the crystals or mineral particles of which the rock is composed.

The microscope, furthermore, affords the means of extending these investigations by enabling the observer to see structural peculiarities which unassisted vision fails to detect.

The following classification has been framed for the purpose of bringing certain important typical rocks prominently before the student's notice, these type-rocks constituting, as it were, the nuclei of their respective groups. Since the groups of each class merge into one another more or less in mineral constitution, no sharp boundary lines can be drawn between them; the type-rocks of the different groups therefore serve as milestones by means of which the student may ascertain in what part of the great series to class any particular rock; the types holding a relation to the whole petrological series somewhat analogous to that which Frauenhofer's lines bear to the spectrum.

CLASSIFICATION OF ROCKS.

Eruptive Rocks.

I. *Vitreous.*

Obsidian
Pumice } including hyaline rhyolite.
Perlite
Pitchstone
Tachylyte.

II. *Crystalline.*

A. Typical groups
- Granite group
- Felstone „
- Syenite „
- Trachyte „ including rhyolite proper.
- Phonolite „
- Andesite „
- Porphyrite „
- Diorite „ ⎫
- Diabase „ ⎬ included under the old term 'greenstone' in its original and broadest signification.
- Gabbro „ ⎪
- Basalt „ ⎭

B. Rocks of exceptional mineral constitution.

III. *Volcanic Ejectamenta.*

IV. *Altered Eruptive Rocks.*

SEDIMENTARY ROCKS.

I. *Normal Series.*

- Arenaceous group (sands).
- Argillaceous „ (clays).
- Calcareous „ (limestones).

II. *Altered Series.*

A. With no apparent crystallisation.

B. With sporadic crystallisation.

C. Crystalline
 - *a.* non-foliated.
 - *b.* foliated and schistose.

III. *Coarse Fragmental Series.*

Breccias and conglomerates.

IV. *Tufas and Sinters.*

V. *Mineral Deposits constituting Rock-Masses.*

ERUPTIVE ROCKS.

Class I.—Vitreous Rocks.

The vitreous rocks are characterised by their generally homogeneous aspect, their more or less glassy lustre (which, however, is sometimes only feebly glassy, greasy, or dull when the rock is partially devitrified), by their conchoidal fracture, and, optically, by the single refraction which they exhibit, except when more or less crystalline structure has supervened. They may, like the crystalline eruptive rocks, be divided into two sub-classes, the highly-silicated or acid (those containing over 60 per cent. of silica), and the basic (or those which contain less than 60 per cent.). Some of those usually occurring in the former sub-class vary somewhat in the amount of silica which they contain, and at times appear to belong rather to the basic sub-class; the pumice from some localities, for example, having less than 50 per cent. of silica, while that from others contains over 62 per cent.

The vitreous rocks may be conveniently arranged in the following order:—

I. Containing over 60 per cent. SiO_2:—

>Obsidian.
>Pitchstone.   Pumice.
>Perlite.

II. Containing less than 60 per cent. SiO_2:—

>Tachylyte.   Pumice.

The vitreous rocks of the first or highly-silicated sub-class closely resemble the liparites, trachytes, andesites, and other highly-silicated eruptive rocks in their chemical composition, while the minerals which are developed in many of them also imply a similarly close relationship. So close, indeed, is this relation that some petrologists include obsidian, pitchstone, perlite, pumice, and certain quartziferous trachytic lavas, under the terms rhyolite and liparite.[1] The student should therefore bear in mind the fact that the separation of the vitreous from the crystalline rocks refers merely to physical differences which the members of these two sub-classes respectively present, and does not imply any special difference in their chemical composition. These physical differences depend upon the conditions under which solidification was effected, whether gradual or rapid. In the former case the molten mass would develop crystals, in the latter it would remain amorphous: it would, in fact, result in a more or less perfect glass. In these natural glasses it is, however, common to find crystallites and crystals, the former usually developed very completely, the latter less perfectly formed as a rule, since they generally present rounded boundaries, or their angles, if any exist, also appear rounded. The cause of these rounded boundaries does not, as yet, seem to be satisfactorily determined. It is known that fragments of rock and individual crystals become rounded by constant attrition during their ejection from, and their returning fall into, the throat of a volcano; and since, in rather rare instances, the microscope shows that some of these vitreous lavas contain not merely rounded crystals, but also well rounded fragments of other vitreous rocks of a quite distinct and different character from that of the matrix in which they are enveloped, it seems possible that in such cases the rounded crystals and fragments of rock represent ejectamenta, which, rounded by attrition, and lying within or around the crater, have been taken up by the viscous mass of lava as it welled over them. If this were the case, we might at first be tempted to think that the rounding was due to the superficial fusion of the

---

[1] The name rhyolite, from ῥύαξ (a lava stream) and λίθος, was introduced by v. Richthofen in 1860, and included certain Hungarian quartz-trachytes, which showed strong evidence of viscous fluxion, and the highly silicated vitreous rocks just mentioned. A year later Justus Roth applied the term liparite to similar crystalline and vitreous rocks occurring in the Lipari Islands.

fragments or crystals, just as fragments of minerals become fused in a borax bead before the blowpipe; and it may be that such a supposition is not wholly incorrect, since, although in the borax bead the minute fragment as it fuses becomes surrounded by visible tortuously-twirling strings of its own molten substance, before these fused products become perfectly incorporated with the borax glass; still we must remember that this fused, ropy matter is visible, because it differs in density from the fused borax, while in the case of vitreous rock fragments, felspar crystals, &c., fusing in a highly heated vitreous magma, the respective specific gravities do not differ sufficiently to render the phenomenon of imperfect incorporation apparent.

The sp. gr. of borax	is	1·71
,, obsidian	,,	2·41 – 2·57
,, perlite	,,	2·25
,, sanidine	,,	2·56 – 2·6
,, plagioclase	,,	2·56 – 2·76

It should, however, be borne in mind that the substance of felspars, which are the principal rounded crystals in vitreous rocks, is approximately colourless, so that in a colourless magma the phenomena of imperfect mixture would not be apparent. Such phenomena are, however, distinctly visible in some obsidians and pitchstones, in which, under the microscope, tortuous lines of glass, differing markedly in colour or tint from the glass in which they lie (fig. 75), denote, no doubt, a difference in the relative specific gravities of the two glasses.[1] Such included glass lines and bands in hyaline rhyolites, although they show us that the glass is not homogeneous, do not furnish us with any clue as to the source of the material which differs from its matrix. It cannot well be imagined that the rounded crystals and

[1] This may be seen in the obsidians from Tolcsva in Hungary, Truckee Ferry in Nevada, and other similar rocks. Kindred phenomena may be seen on mixing liquids of different specific gravities.

fragments in these vitreous lavas were showered down on the surface of the viscous lava stream, since that would imply a synchronous eruption of lava and ejection of ashes, dust, sand, &c., from the same vent, for, where two craters are situated near one another, one is generally at rest while the other is active. If neither of the foregoing hypotheses be adopted there seemingly remains but one other, namely, that the crystals have been developed during the solidification of the rock, and that the rounded contours which their sections present are due to aborted crystallisation, such as that pointed out by Poussin and Rénard as occurring in the orthoclase of the porphyrite of Mairus in Belgium.[1] We may, perhaps, admit with safety that in many instances the crystals have been developed in the rock; but, if we admit it in all cases, how are we to account for the included fragments of rock which may occasionally be noticed in microscopic sections of these lavas? It is also worthy of note that the same section may exhibit crystals with well-developed angles and also rounded crystals of the same mineral.

Certain structural peculiarities and inclosures, many of which can only be observed microscopically, are characteristic of the vitreous rocks. These structures or inclosures do not always individually characterise these rocks, since it is not uncommon to find crystals, crystallites, microliths, spherulites, &c., all developed in the same specimen.

The following is a descriptive list of the principal structures which occur in vitreous rocks.

Homogeneous.—This condition is more hypothetical than real, since, when examined microscopically, scarcely any of the most homogeneous-looking obsidians are seen to be free from inclosures of microliths. If, however, these microliths and other inclosures be put out of the question, the glassy

---

[1] *Mémoire sur les caractères minéralogiques et stratigraphiques des Roches dites Plutoniennes de la Belgique et de l'Ardenne Française.* De la Vallée Poussin et Rénard. Bruxelles, 1876.

Banded and Damascened Structures.

matrix in which they lie may be regarded as homogeneous, or as approximately homogeneous, although, under high powers, it often shows included dusty matter, which might exhibit some definite characters if still higher powers were employed. It may, however, be stated that, as a rule, all of these natural glasses contain fine dust and microliths.

Banded.—This structure is rendered evident by the interlamination of glasses which differ in tint, or by the segregation of granular matter in strings or layers. The bands are seldom continuous for any distance, and usually exist merely as elongated lenticular streaks. Fig. 74 A shows the banded appearance of a section of black obsidian from the Island of Ascension, magnified 50 diameters.

FIG. 74 A.

Damascened.—The author suggests this term as a convenient one by which to describe the structure shown in some obsidians, in which streaks or threads of glass are contorted in a confused manner, which somewhat resembles the markings on Damascus sword-blades or the damascening on gun-barrels. Fig. 75 represents part of a section of a red obsidian, from Tolcsva, in Hungary, magnified 50 diameters, in which the damascene structure is well shown. These twisted threads of glass are of a different tint or colour to that of the glass in which they lie. The appearance which they present when seen in thin section under the microscope suggests that which two liquids of different density exhibit when they are imperfectly mixed and slightly agitated, as pointed out on page 179.

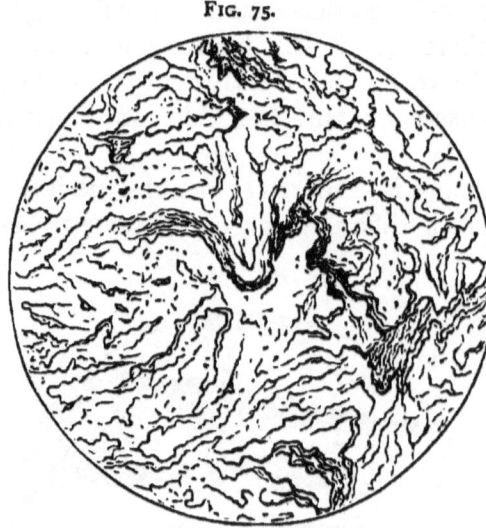

Fig. 75.

Perlitic.—A structure especially characteristic of the rocks termed perlites, but sometimes developed in other vitreous rocks such as trachylyte, &c. This structure has been described as a phenomenon attendant upon contraction, first by Professor Bonney and subsequently by the author, who was at the time ignorant of Professor Bonney's conclusions. These conclusions have since been admirably demonstrated by Mr. Allport's examination of some ancient perlites occurring in Shropshire. The structure consists in the development of numerous minute cracks which exhibit varying curvature, and produce somewhat concentric and approximately spheroidal or elliptical figures, but the lines which bound these forms do not coalesce as a rule, so that the structure may be described as an imperfect, concentric, shaly one, which, on a large scale, finds a parallel in the spheroidal structure developed in some basalts. The spheroids in perlite are almost invariably found to lie packed between minute rectilinear fissures which traverse the rock in all

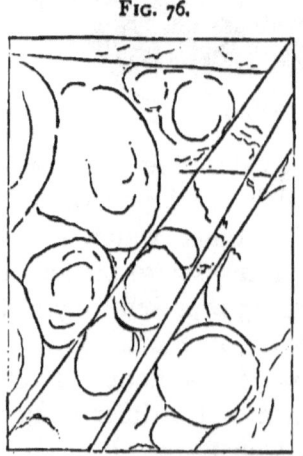

Fig. 76.

directions, but which are seldom seen to cut through the spheroids. (Fig. 76 represents a section of perlite from Buschbad, near Meissen, Saxony, magnified about 10 diameters.) The spheroidal bodies are often seen to be traversed by more or less parallel streams of microliths which bear no relation, or observe no relative disposition, to the spheroids, thus showing that the perlitic structure had no existence when the rock was in a state of fluxion, but was developed on the solidification of the rock.

Spherulitic.—This is a structure totally distinct from that just described and may be regarded as concretionary, or as resulting from incipient crystallisation around certain points or nuclei. The nuclei, when they exist, consist either of a granule or a minute crystal or crystallite, but most commonly no nucleus is discernible. Spherulitic structure in its most rudimentary form seems to consist in the segregation, in spots, of glassy matter, of a different colour to that which constitutes the matrix, and often contains a considerable quantity of very fine dust, the nature of which has not been ascertained but which is probably magnetite. The glass constituting the spherulites is usually of a deep yellowish-brown colour and, in very perfectly developed spherules, generally forms a broad zone, within which lies clear light-coloured or colourless glassy matter having a radiate structure, due to imperfect crystallisation, while, at the central spot, from which the crystals, rods, or fibres emanate, a few doubly refracting granules may sometimes be observed. In some instances, as in the obsidians of the Lipari Islands, a per-

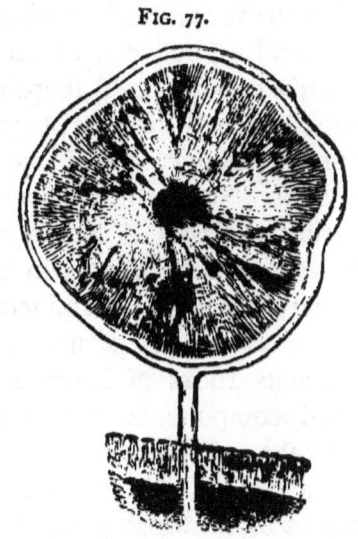

FIG. 77.

fectly clear, colourless, but very narrow outer zone surrounds the brown glassy envelope of the spherulite, as in fig. 77. (Magnified 150 to 200 diameters). In vitreous rocks from the last named locality, in those from the Island of Ascension, and in many other examples, the spherulites occur in definite layers or belts, and have, in many cases, been elongated in the direction in which the lava-stream flowed; at times they even coalesce and form more or less continuous bands, as in fig. 78 (magnified 22 diameters), which represents part of a band of coalesced spherulites in the obsidian of Rocche Rosse, Lipari. Occasionally, but very rarely, spherulitic structure is so extensively developed in vitreous rocks that the whole mass consists of closely packed spherulites, between which only small patches of the glassy matrix can here and there be discerned, while the spherulites are so closely crowded together that their boundaries are no longer spherical, but, by compression, assume polygonal forms. Spherulitic structure is sometimes developed in artificial glass. A fragment of a plate-glass window, from a house which had been burnt down, exhibited colonies of spherulites, when examined under the microscope.

FIG. 78.

Axiolitic.—A structure is developed in some of the vitreous rocks of Nevada, U.S., and elsewhere, the individual components of which have been termed axiolites by Zirkel.[1] These appear to be somewhat analogous in structure to spherulites; elongated lenticular and curved zones of brownish glass forming the envelope of a smaller corresponding mass of paler vitreous matter, in which

---

[1] *Microscopic Petrography*, Zirkel, U. S. Exploration of the Fortieth Parallel.

incipient crystallisation or fibrous structure trends at right angles to the inner surfaces of the envelope towards a longitudinal median line. The great diversity exhibited by such structures is well shown in the work cited in the footnote from which the accompanying figure (79) is copied. The figure represents the axiolitic structure visible in a rhyolite from N.W. of Wadsworth, Nevada, U.S.

FIG. 79.

Porphyritic.—This term is applied to vitreous, just as to other rocks, implying that isolated crystals distinctly visible to the naked eye occur in them. The application of this term has in all cases a purely arbitrary limit, since it not merely refers to the mode of occurrence of the crystals, but also to their size; rocks in which very small isolated crystals occur only being spoken of as micro-porphyritic, simply because, from their small dimensions, they do not convey to the naked eye the blotched appearance which characterises the commonly recognised porphyries.

Trichitic and *Microlitic* are terms which might also be given to those vitreous rocks which contain multitudes of the bodies already described as trichrites and microliths; but as nearly all vitreous rocks are more or less microlitic, and as the word 'trichitic' sounds inconveniently like the adjective 'trachytic,' which latter is often applied to rocks of this class, such terms as trichitic and microlitic are perhaps better left alone.

Devitrified.—This implies that the rock has undergone, to a greater or less extent, certain physical changes which cause it no longer to behave as a glass, its vitreous character being partially or completely destroyed by the development either of microliths, crystalline granules, or crystals. The ultimate stage of crystalline-granular devitrification is felsitic matter, and, when a rock has undergone complete change of this kind, it is only possible to arrive at conclusions as to its once vitreous nature, by means of those structural peculiarities which indicate former fluxion, and, should those characters fail to be very well marked, it is, as a rule, most hazardous to jump at any conclusions concerning the original condition of the rock.

Filiform.—A condition occasionally, but rarely, met with ; as in the filiform lava of Hawaii, in the Sandwich Islands, known as Pélé's hair, in which molten vitreous lava has been frayed out and blown by the wind into long and extremely slender glassy threads, which commonly terminate in little fused knobs or pellets. This structure is also produced artificially in blast-furnace slags.

OBSIDIAN.

Obsidian results from the quick solidification of lavas which, if slowly cooled, would develope crystalline structure and assume the character of trachyte, liparite, &c., rocks which contain over 60 per cent. of silica. In obsidian no crystalline structure is developed; it is a true, natural glass. Nevertheless, obsidians frequently contain microliths; and, when spherulitic, the spherulites commonly show a radial crystalline or fibrous structure. Obsidians present a very homogeneous appearance and a strong vitreous lustre. Their fracture is eminently conchoidal. They vary somewhat in colour, but are mostly black or grey. In thin splinters they are all more or less transparent.

Obsidians also vary in chemical composition. The silica may be estimated at from 60 to 80 per cent., the alumina at 18 to 19 per cent., while the remaining constituents are potash or soda, lime, magnesia, peroxide of iron, and occasionally as much as 0·5 of water. Their specific gravity ranges between 2·4 and 2·5, and the hardness equals 6 to 7.

Those which contain the largest proportion of silica are only slightly acted upon by acids.

Before the blowpipe obsidian is fusible on the edges of thin splinters. Sections of obsidian when placed under the microscope between crossed Nicols exhibit no double refraction, the field appearing quite dark; but this dark field is usually thickly studded with bright doubly-refracting microliths, and under moderately high powers crystallites of varied forms, exhibiting structural peculiarities of excessive beauty and interest, may often be met with in great profusion.

There are, however, some obsidians, such as the pseudo-chrysolite or bouteillenstein, which occurs as rounded pebbles in sand at Moldauthein in Bohemia, and in some of the tuffs near Mont Dore in Auvergne, which show no crystallites under the microscope, and equally pure obsidians occur in one or two localities in New Zealand and in Iceland. All of these, however, contain great numbers of gas pores.

The crystallites which occur in obsidian vary so greatly in form that mere descriptions of them would be of little use to the student. The precise mineral species which they represent are in many cases undetermined, but it is probable that many of them are incipient felspar crystals. In some of the small crystals or microliths, which are so common in rocks of this class, it is interesting to note the gradual development of structure which may sometimes be seen in a single microscopic section; in one place simply a comb-like crystalline growth springing from minute tapering rods, which constitute, as it were, the visible axes of these little crystallites; in another, a microlith, or small crystal, in which may be seen a structure identical with the preceding, and which seems to show the plan upon which it has been built up, to be in fact the framework upon which it has been developed. The little axial rods, if they may be so termed, are not always straight; at times they have a somewhat sigmoidal flexure, at others they occur in pairs arranged like two bows set back to back. This disposition is occasionally coupled with the intermediate development of a small, square, rectangular, or rhomboidal mass, from the four corners of which the apparent homologues of these arcuate rods sprout out like horns, as in fig. 80, while the whole is surrounded by a hyaline border, whose external boundary and occasionally striated structure indicate differentiation of the surrounding glassy magma and the incipient extension of crystalline development. Similar borders often surround crystallites which give sections like those which an elongated pyramid would afford; and cruciform groupings, which closely resemble aggregates of such pyramids, may also be seen at times in obsidians.

FIG. 80. ×350.

Besides the crystallites just mentioned, it is common to find spherulites developed in these lavas. In their most rudimentary condition they occasionally seem to be represented by blotches of a glass, of deeper colour than that of the surrounding matrix. Generally the most perfectly developed spherulites have a somewhat broad border of glass, which appears reddish-brown by transmitted light, and surrounds a central spherule of clear and often almost colourless glass, in which a radiate structure is developed, while, in some instances, the whole spherulite is surrounded by a narrow colourless envelope of clear glass. These spherulites are sometimes elongated in the direction in which the once viscous stream of obsidian flowed, and this elongation has occasionally taken place to such an extent that the spherulites have coalesced, and formed more or less continuous bands, of which the central portion consists of vitreous and, at times, micro-crystalline matter. This is cased in an outer envelope of glassy matter which appears reddish-brown by transmitted light, and generally snow-white or greyish by reflected illumination. Occasionally this is surrounded by a thin external coat of clear, colourless glass, which, unlike the clear absorption areas seen around the crystallites and dust segregations in some vitreous rocks, is bounded by a

Obsidian.

sharp line of demarcation from the glass which constitutes the matrix. Such spherulitic bands have sometimes, when the coalescence of the spherulites has only extended to their cortical zones, merely the aspect of beads closely strung together ; but, in cases where the coalescence has been more complete, the boundaries of the bands are approximately parallel straight lines, so that the structure of such a band or string may be diagrammatically represented as in fig. 81, *a* being the transverse and *b* the longitudinal section. It is not, however, to be supposed that, where vast multitudes of spherulites are developed on the same plane, transverse sections such as *a* (fig. 81) are invariably to be procured, and in such cases we may assume that the coalescence of the spherulites gives rise to sheets, rather than strings, the vertical sections through such sheets affording in all directions a disposition corresponding with *b* (fig. 81). In some cases, as in the spherulitic obsidian of Rocche Rosse in the Island of Lipari, clear colourless rods of glass are seen to have been extruded through the cortical layers of the spherulitic bands into the surrounding glassy matrix (fig. 82), which are further enlargements of the little rods shown in fig. 78, page 184.

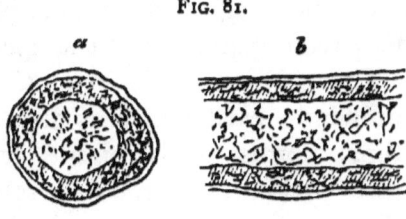

FIG. 81.

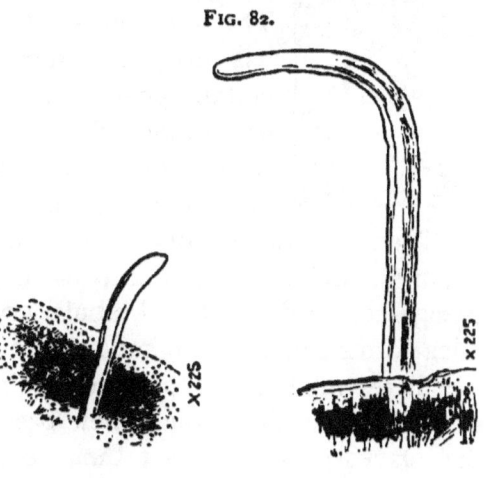

FIG. 82.

After emergence from the band they are frequently hooked

or bent, but not, as a rule, in any mutually definite direction, and in most sections they are seen either to terminate blindly in rounded ends, or to be cut off on the ground surfaces of the preparation.

Small crystals and microliths, as already stated, are of common occurrence in some obsidians. In many cases they can be safely identified with recognised minerals, such as sanidine, plagioclase, augite, hornblende, olivine, tourmaline, zircon, magnesian mica, specular iron, and magnetite.

The felspars occur in small prisms. The augite and hornblende exist either as distinct crystals, similar to the ordinary forms, or as minute acicular bodies and spicular forms ('belonites') which are often bordered by imperfectly radiate, fibrous or hazy, and almost dendritic tufts, which cause them somewhat to resemble the fronds of ferns. Beautiful crystallites of this description may be seen in some of the pitchstones of Arran (fig. 83), and have been identified by S. Allport as augite. The crystals of olivine are always of moderate size and no microliths of this mineral have as yet been detected. The occurrence of tourmaline and zircon has not been definitely determined, but certain prisms belonging to the hexagonal or rhombohedral system have been thought to be tourmaline, while some tetragonal forms are regarded as zircon.

FIG. 83.

The magnetite occurs in opaque octahedral crystals or granules, and the specular iron in little yellowish-red or orange-coloured hexagonal tabular crystals. The magnesian mica forms small, deep, reddish-brown scales and crystals which resemble those of specular iron, but although they show no dichroism when their basal planes coincide with the planes of section, yet their sections are very strongly dichroic when cut transversely or obliquely to the basis.

Obsidian occurs on a large scale as lava-flows, which

frequently present very rough and jagged surfaces. That forming the Rocche Rosse in the island of Lipari is a good example. The accompanying figure of this obsidian stream, which has issued from the crater of Campo Bianco, breaching one side of the crater, and flowing over the white pumice tuffs, of which the cone is composed, is copied by permission of Prof. J. W. Judd from one of the sketches published in his 'Contributions to the Study of Volcanoes,' *Geological Magazine*, Decade II. Vol. ii. No. 2, p. 66.

FIG. 83 A.

Obsidian at times becomes vesicular. In some of the obsidians of Hawaii the vesicles are quite spherical; in others they are elongated or otherwise distorted. Occasionally these rocks are very finely vesicular, as is the case with some of the obsidians of the Lipari Isles, and the vesicles are at times so numerous that the rock acquires quite a frothy character, and passes into pumice.

Obsidians occur in districts where trachytic rocks are common.

PUMICE.

Pumice is a porous, vesicular glass; the vesicles being frequently elongated, sometimes in a more or less definite direction, while at others they anastomose, and give rise to an irregular network of fibrous, intervesicular matter.

Pumice varies considerably in chemical composition, the percentage of silica ranging between 57 and 73. The alumina varies from 9 to 20 per cent., and the remainder consists of lime, magnesia, potash, soda, and peroxide of iron. Water is also present. The specific gravity of pumice varies from 1·9 to 2·5.

The fusibility before the blowpipe is greater in some specimens than in others.

When examined microscopically some pumice appears to consist of interlacing or anastomosing vitreous fibres, few or no microliths or crystallites being developed in the glassy matter. In others microliths and small crystals occur in abundance, and frequently show the stream-like disposition which results from fluxion, or from the drawing out of small portions of viscous lava. The microliths, when they can be determined, are found in most instances to be felspars, both monoclinic and triclinic. Magnetite is also of common occurrence.

Pumice is developed on the surfaces of obsidian streams, and in such cases can only be regarded as a highly vesicular, spongy, or fibrous condition of obsidian. The porphyritic development of felspar crystals in pumice begets the rock termed trachyte-pumice. Pumice also occurs in the form of loose ejected blocks and fragments. These ejectamenta sometimes constitute volcanic cones. In fig. 83 A the cone, which is partially broken down by the stream of obsidian, consists of pumice fragments.

PERLITE (*Pearlite, Pearlstone*).

Perlite is sometimes quite glassy in appearance, but it more frequently exhibits a shimmering, pearly, enamel-like, or greasy aspect on recently fractured surfaces. The colour is mostly pale-greyish, bluish-grey, or yellow-brown. It often appears to consist in great part of spherical or roundish grains, which have a somewhat concentric shaly structure.

In chemical composition the perlites approximate to the quartz-trachytes (the rhyolites proper, or the liparites of Von Richthofen[1]). They contain from 70 to over 80 per cent. of silica. When heated they give off water, the amount varying from 2 to 4 per cent. Their hardness is about 6.

Perlite must be regarded as the vitreous condition of the felsitic rhyolites, and, like other vitreous rocks, plays a very subordinate part in the constitution of the hitherto explored portions of the earth's crust, when compared with those rocks of which it is the vitreous representative. It should, however, be borne in mind that the originally vitreous character of many eruptive rocks has yet to be discovered, since, owing to devitrification, they frequently present appearances which, in the absence of microscopic investigation, afford no clue to the physical characters which they possessed at the time of their eruption. It is indeed more than probable that many of the so-called hornstones, felstones, and even rocks, which were mapped by the older geologists as greenstones, are merely rhyolitic rocks in a devitrified condition. In this country the researches of Professors Bonney and Judd, Mr. Allport, and the author, have already demonstrated the existence of rocks of a rhyolitic type, of which the real characters had previously been overlooked. When examined microscopically thin sections of perlite exhibit numerous fissures, and between these fissures great numbers of somewhat concentric cracks are visible, causing a separation of the rock into more or less regular spheroids. The cracks do not appear to join up in continuous ellipses, but

[1] The term liparite has been applied by Roth to the whole of the rhyolites. Von Richthofen, however, limits the use of the term to the felsitic rhyolites, or rhyolites proper.

thin off and at times overlap. They nevertheless form approximately concentric envelopes around the spheroidal nucleus of glass. It is worthy of remark that these bodies are seldom or never traversed by the straight cracks which run in various directions through the rock, but lie between them, often closely packed, distorted, and apparently compressed against the planes of the straight fissures. This indicates that the straight fissures were formed first and that the spheroidal or perlitic structure, as it may be termed, was subsequently developed.

Streams of microliths commonly occur in these rocks, and they traverse the perlitic bodies without the slightest indication of deflection. They have, in fact, been quite uninfluenced in their direction of flow by the minute structural planes and elliptical cracks which occur so plentifully in the rock.

Crystals, both megascopic and microscopic, occur in considerable numbers in some perlites. They consist principally of sanidine and plagioclastic felspars, magnesian mica, magnetite, and occasionally specular iron. The microscopic inclosures consist mostly of felspar-microliths, trichites, and belonites. Spherulitic structures are also present at times.

That the perlitic structure has probably resulted from the development of more or less concentric zones of contraction on cooling has been pointed out, both by Professor Bonney, and by the author. Perlitic structure bears a somewhat close relation to the larger spheroidal structure which is occasionally to be seen in basalt.

Indications of perlitic structure may be observed in thin sections of vitreous rocks other than perlite, and the author has noted the incipient development of this structure in an Irish tachylyte (Journ. Royal Geological Society, Ireland, vol. iv., p. 230), thus showing that the structure, although characteristic of some highly silicated vitreous rocks, is not exclusively peculiar to them.

Professor A. von Lasaulx points out the fact[1] that many spherulitic rocks were formerly regarded as perlites, but he adds that spherulitic and perlitic structures are totally different, since the latter consist merely of little masses of glass, while the former are 'crystalline individualisations.'

Spherulites may in fact be regarded as spots of devitrification, while perlitic structure is simply a phase of fission resulting from contraction on cooling, the glass included by the elliptical fissures in no way differing from the surrounding glass.

Pitchstone.—The pitchstones may be regarded as vitreous conditions of trachyte on the one hand, and, on the other, of those rocks which range from granites to felstones, including the porphyritic varieties of those rocks such as quartz-porphyry, porphyritic felstone, &c. It is true that pitchstone is not recognised as a vitreous condition of granite, but since passages are known from granite into quartz-porphyry, and from quartz-porphyry into felstone, it seems that we may fairly be allowed to regard the vitreous equivalent of the felstones and quartz-porphyries as, at all events, an indirect vitreous phase of granite, which appears more directly to find its rhyolitic representative in the nevadites. The pitchstones are classed as trachytic and felsitic. The devitrification of the trachytic pitchstones is effected by the development of microliths, which for the most part consist of sanidine and hornblende, while the felsitic pitchstones become devitrified by the setting up of a micro-felsitic or, as v. Lasaulx terms it, a micro-aphanitic structure.

The pitchstones have, as a rule, a perfectly conchoidal fracture, sometimes rather splintery, and a more or less greasy semi-vitreous, or pitch-like lustre, whence the name. They are mostly blackish-green, dark olive green, or brown. Occasionally they are red or dull yellow.

---

[1] *Elemente der Petrographie*, p. 223. Bonn, 1875.

In chemical composition they resemble obsidian, but contain from four to ten per cent. of water. If, however, the water be omitted, and the other constituents brought up to 100, the composition of pitchstone may be regarded as approximately identical with that of trachyte. The silica ranges from 63 to 73 per cent., the alumina from 9 to 13, and the alkalies from 2 to 8 per cent. Before the blowpipe some varieties do not fuse so readily as others, but they all melt either to a frothy glass or to a greenish or greyish enamel, and yield water when heated in a tube.

The pitchstones have a hardness of 5 to 6, and a specific gravity of 2·2 to 2·4. They are not acted upon by acids.

The pitchstones, as already stated, may be divided into two groups—the trachytic pitchstones and the felsitic pitchstones. The former are related both geologically and in mineral constitution to the liparites, while the latter are related to the quartz-porphyries and felstones. With regard to the trachytic pitchstones, Rosenbusch observes[1] that no sharp line of demarcation can be drawn between them and the glass-magma-liparites which contain water.

Sections of pitchstone, when examined microscopically, appear, like other glasses and amorphous substances, dark between crossed Nicols. By ordinary illumination they appear either as perfectly homogeneous glass containing crystals and microliths, or they exhibit streaks of glass of a deeper or paler tint than that of the surrounding matrix, which in their disposition at once suggest the presence of fluxion structure. In the latter cases, as well as in the former, crystals are of common occurrence, and microliths are usually developed in abundance.

The crystals are mostly sanidine, hornblende, and magnetite, but plagioclase is not uncommon.

The microliths may often be recognised as consisting of the above-mentioned minerals, and in some pitchstones, especially in those from the Isle of Arran, microliths of

[1] *Mik. Phys. d. Massigen Gesteine*, p. 160.

augite are also plentiful. The pitchstone of Corriegills in Arran shows these augite microliths or belonites in great profusion, forming more or less stellate groups of pale greenish forms, which somewhat resemble the fronds of ferns, and which have been admirably figured and described by Zirkel, Allport, Vogelsang, and other microscopists. Fig. 84 represents a magnified section of Arran pitchstone, and is copied, by permission of Prof. Zirkel, from his 'Mikroskopische Beschaffenheit der Mineralien u. Gesteine.'

FIG. 84

The glass of most pitchstones appears, when light is transmitted through thin sections, of a pale yellowish, greenish, or brownish tint, but it is scarcely safe to venture a decided opinion upon the precise nature of the pigment. Sometimes, but not in all cases, there appears to have been an abstraction of this pigment from those portions of the glassy magma which immediately surround the microliths of magnetite and augite, this absorption of pigment producing a comparatively clear, colourless ring around these bodies. Steam pores, and occasionally spherulitic structures, are met with in pitchstones. The microliths in these rocks usually lie in streams, which sweep round the larger imbedded crystals, and indicate, in a marked manner, the originally viscid condition of the matrix in which they occur.

The felsitic pitchstones bear much the same relation to the felstones that the trachytic pitchstones bear to the trachytes, while the porphyritic varieties, termed pitchstone porphyry, may also be regarded as the vitreous equivalent of quartz-porphyry. To these pitchstone porphyries Vogelsang has given the name vitrophyre.

The principal porphyritic crystals in these rocks are quartz, sanidine, plagioclase, augite, hornblende, magnetite, and biotite. They also at times contain inclosures of glass, which are sometimes devitrified by conversion into micro-felsitic matter, at other times, by the development of microliths.

The devitrification of the magma of the felsitic pitchstones is as a rule micro-felsitic, while that of the trachytic pitchstone is effected by the development of microliths. Microliths, some doubly refracting and some singly refracting, occur however in the glasses of nearly, if not all different varieties of pitchstone. Inclosures of fluid, containing mobile bubbles, are of rare occurrence in these rocks.

The devitrification of the porphyritic and felsitic pitchstones does not, as a rule, take place uniformly, but occurs in a seemingly capricious manner, often being developed in irregularly-distributed and irregularly-shaped patches, sometimes occurring in spots, sometimes in strings, which usually indicate fluxion-structure. Dark granules also occur in these rocks, which possibly represent magnetite, and these granules frequently assume a string-like or banded arrangement, while irregular strings or lenticular flecks of included glass also denote a former state of fluxion.

In their most perfectly vitreous conditions it seems that no sharp line of demarcation can be drawn between the different varieties of pitchstone here described, for although differences may, and doubtless do exist, still our power of appreciating those differences is limited or *nil*, so long as devitrification has not supervened, and so indicated, by incipient crystalline development, the true petrological affinities of the glass. The crystals which occur porphyritically in the different varieties afford us a very imperfect clue to these relations, simply because they represent for the most part identical mineral species. The characters of the rocks with which they are associated, and of which they represent the vitreous conditions, give us however a more exact notion of the places which they ought to occupy in our classification of them.

Indeed it may not be indiscreet to believe that in many, if not in all instances, the crystalline equivalents of these vitreous rocks do but represent an advanced phase of devitrification, and that all trachytes were once trachytic pitchstones, and that all felstones were once felsitic pitchstones, either at the time of or prior to their eruption.

Tachylyte.—The rocks included under this name must be regarded as vitreous conditions of basic rocks, especially of basalts. These glassy basalts are termed basaltvitrophyres by Rosenbusch, and he subdivides them into tachylytes, or those which are soluble in acids, and hyalomelanes or those which are insoluble in acids. The tachylytes occur mostly as *salbands*, or thin crusts at the sides or margins of basalt dykes, but the essentially vitreous basic lavas, such as those of Kilauea in the Sandwich Islands, which form actual flows of considerable magnitude, constitute, as pointed out by Cohen, independent rock masses. These Kilauea lavas are, as a rule, rich in olivine, and are for the most part highly vesicular. Tachylyte also occurs lining or filling vesicles or cavities in basalt. The tachylytes are black or brown glasses and somewhat resemble obsidian, but when struck with the hammer they do not usually afford large flakes and extensive conchoidal fractures like obsidian, but generally break up into small irregular fragments and splinters. They also differ from obsidian in point of fusibility and chemical composition, tachylyte only containing from 50 to 55 per cent. of silica. The remaining constituents are alumina, protoxide of iron, sometimes peroxide of iron, lime, magnesia, potash, soda, and usually about 6 or 7 per cent. of water. From the few analyses of tachylyte which have hitherto been made, the composition appears to be rather variable, but the percentage of silica is pretty constant. Before the blowpipe tachylyte fuses quite easily [1] with intumescence to a dark slaggy glass. It is decomposed with gelatinisation in hydrochloric acid. Its hardness is about 6·5, and its specific gravity about 2·5.

Under the microscope tachylytes vary greatly in appearance, some being comparatively translucent, at all events in places, while others are almost wholly opaque, even in excessively thin sections.

This opacity seems in many cases to be due to the presence of fine opaque, black, dusty matter which pervades a great portion of the glass, but is more densely segregated around certain spots these spots being frequently small crystals of magnetite. Fig. 85 represents part of a section of tachylyte from Slievenalargy, Co. Down, Ireland (magnified 300 diameters), in which these dust accumulations are well shown. Where these dust segregations are less dense, the sections often appear of a brownish colour, and this is possibly due to peroxidation, the magnetite probably being converted into martite, or some closely allied mineral. In the clearer portions of the same section from which fig. 85 was drawn, numerous opaque patches of irregular form are visible (fig. 86, magnified 55 diameters); their boundaries are sharply defined, while they

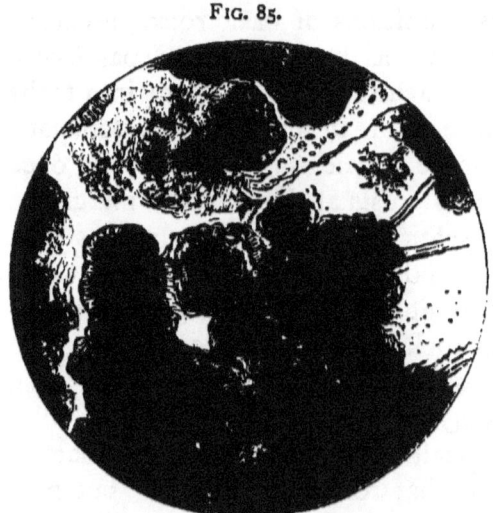

FIG. 85.

---

[1] Whence the name from ταχύs, quickly, and λυτόs, fusible.

are bordered by the clear absorption spaces from which the dust, so finely disseminated through the other portions of the glass, appears to have been abstracted. At A A in the same drawing lacunæ of greenish coloured ferruginous glass are shown, which mineralogically are possibly allied to glauconite. At B B B portions of fine rod-like bodies are delineated, and

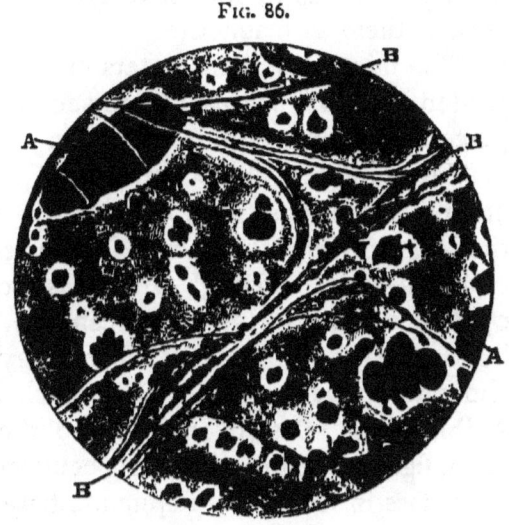

Fig. 86.

these also contain, or are fringed with, green matter. In some places, as in the central part of the figure, a tendency to perlitic structure may be detected.[1] A tachylyte from Bobenhausen is described by Vogelsang[2] as a brownish-red glass, containing dark crystallites which are developed in fern-like forms, but which in most parts of his drawing appear to be slightly-fringed spherical or cruciform bodies, or irregular dark patches. He did not regard this substance as magnetite, arguing that, when magnified 800 to 1,000 diameters, the granules constituting the thinner portions and edges of these crystallites do not appear opaque like magnetite, but somewhat translucent and of a brownish colour. The glass around them appears much clearer and of a yellow colour. Vogelsang, Zirkel, and Möhl all concur in regarding these bodies as similar to those which often occur in

---

[1] Fuller details respecting this rock will be found in a paper by the author 'On Microscopic Structures in Tachylyte from Slievenalargy, Co. Down,' *Journ. Royal Geol. Soc. Ireland*, vol. iv. part 4, new series, p. 227.

[2] *Die Krystalliten.* Bonn, 1875, p. 111.

blast-furnace slags. Both Vogelsang and Zirkel consider these crystallites to consist of ferruginous glass, while Möhl regards them as magnetite.

The microscopic characters of several tachylytes are recorded by Zirkel in his 'Untersuchungen über die Basaltgesteine,' Bonn, 1870, p. 182. In one from Meinzereichen in Hesse, he cites the occurrence of fern-like developments around a transverse section of an apatite crystal. In a section of tachylyte from Some in the Isle of Mull, the whole of the section, although an excessively thin one, appears opaque or very feebly translucent on the edges, while embedded in it are minute transparent crystals which frequently show hexagonal sections, and which are probably apatite. Möhl, in his 'Basalte und Phonolithe Sachsens,' Dresden, 1873, figures and describes the occurrence of patches of tachylytic glass in sections of nepheline-basalt, which contain fern-like trichites, and occasionally small crystals of nepheline. Rosenbusch describes a tachylyte from Czertochin in Bohemia, as a greenish-grey glass full of strings of minute steam pores which occasionally anastomose. He states that this rock is very quickly and completely dissolved in hydrochloric acid, without the application of heat.[1]

CHAPTER XII.

ERUPTIVE ROCKS.

CLASS II.—CRYSTALLINE ROCKS.

GRANITE GROUP.

Granite.—The granites (from the Latin *granum*, a grain) are essentially, as their name implies, crystalline-granular rocks which may be regarded as consisting typically of

[1] *Mik. Physiog. Min.* p. 139, fig. 15. Stuttgart, 1873.

orthoclase, quartz, and mica. There is, however, considerable variation in the mineral constitution of granites. In some the felspathic component is not merely orthoclase; plagioclastic felspars such as albite and oligoclase being frequently present, while the mica, which is usually muscovite or biotite, may at times be represented by lepidolite, lepidomelane, or other micas. Other minerals, which are not regarded as essential constituents of granite, are often present. When widely disseminated, these accessory constituents play only a very subordinate part; but, in certain limited areas, they are often developed in sufficient quantity to impart a distinctive character to the rock. Thus, for example, schorl frequently occurs in considerable quantity in granitic masses at or near their contact with other rocks, sometimes to such an extent that the term schorlaceous granite is applied to the rock. Apatite and magnetite are also minerals of common occurrence in granites. Epidote and garnets are less common, but are often met with. Pyrites is common in many granites, and it seems doubtful whether, in some cases, it should be regarded as a mineral of secondary origin. Talc, beryl, iolite, andalusite, topaz, cassiterite, and hematite are occasionally met with in granitic rocks. Hornblende is of common occurrence, and, when tolerably plentiful, the rock is then termed hornblendic or syenitic granite. When quartz is absent, or only poorly represented, and the mica is replaced by hornblende, the rock is called syenite.[1] Chlorite, epidote, pinite, and several other products of the alteration of other minerals, are not of unfrequent occurrence in these rocks.

Kaolin very commonly occurs in granites, and results from the decomposition of the felspars. Graphite is also met with at times.

Granite rocks vary very considerably in texture and in structural characters.

[1] It should, however, be remarked that the term syenite, as first employed by Pliny, and as used in most geological works, until within the last few years, implied hornblendic granite, such as that which comes from the quarries of Syene in Egypt.

The granites, as a rule, are either coarsely or finely crystalline-granular in texture, and, when very fine-grained, and the mica is only poorly represented, or totally absent, pass into felstones of variable texture.[1] The granitic rocks are frequently porphyritic. Crystals of orthoclase several inches in length being of common occurrence in some, while, in others, the mica (usually muscovite) forms, by its conspicuous development, the dominant mineral. When mica is scarce, and the rock assumes a felsitic character, it is common to find either orthoclase or quartz porphyritically developed. In the former case the rock would be styled a felspar porphyry, in the latter a quartz-porphyry, or elvan.[2]

In true granites no micro-crystalline or amorphous paste is visible between the crystals and crystalline grains of which the rock is composed. Of these component minerals orthoclase is generally the most plentiful; next follows quartz, and then mica, in the usual, but not the invariable, order of quantitative importance. In their order of solidification, or crystallisation, quartz appears to come last, and, although it sometimes occurs in hexagonal pyramids or in combinations of the pyramid and prism, still its development, as a rule, seems to have been imperfect, and to have resulted mainly in irregularly shaped, angular, crystalline grains. The orthoclase in granites varies in colour. In some it is red, often of a flesh-red or pink tint, in others white, grey, or yellowish. It very commonly occurs in Carlsbad twins. The crystals are often several inches in length, as in some of the Dartmoor granites.

When granites are weathered, the felspar crystals are converted into kaolin and the rock in course of time crumbles away. The kaolin or china-clay which remains

---

[1] Felsitic matter, which constitutes the chief bulk of felstones, is a very finely crystalline-granular, or micro-crystalline, or crypto-crystalline, admixture of orthoclase and quartz.

[2] Elvan is a Cornish name, and is commonly applied by the Cornish miners to most of the dykes which occur in that county, irrespective of their mineral constitution. The term has, however, of late years been restricted to quartz-porphyries.

after the disintegration of granite frequently contains a large proportion of quartz grains, and this decomposed rock is known as china-stone.

The crystals of orthoclase are not always well developed in granites; they sometimes have very irregular contours, and occasionally their angles are rounded. Under the microscope they frequently present a more or less turbid appearance, and this greatly increases in proportion to the stage of decomposition at which the rock has arrived, until they ultimately become completely kaolinised and opaque. They occasionally, but rarely, contain fluid lacunæ. Plagioclastic felspars, either albite or oligoclase, are of frequent occurrence in granites. They usually occur in smaller crystals than the orthoclase. Under the microscope they exhibit, when fresh, the characteristic twinning of plagioclase, but, as decomposition advances, a granulated structure also supervenes, which obliterates this distinctive structure, and renders it impossible to determine whether they were originally monoclinic or triclinic felspars.

The quartz, as already stated, sometimes occurs in well-developed crystals, and sometimes in angular, crystalline grains. The former often exhibit a polysynthetic structure when examined in polarised light. Under the microscope, in thin sections, they appear quite glassy and clear, and are seen to contain numerous fluid lacunæ, which are often so plentiful, as to impart an almost turbid appearance to the crystal or granule. The contained fluid is generally water or aqueous solutions of chlorides and sulphates of sodium, potassium, and calcium. Apatite crystals are also frequently visible in the quartz of granite. The micas in thin sections of granite appear either in well defined crystals, which, when the section is taken parallel to their basal planes, appear as six-sided tables, or in scales of irregular form. The potash micas appear clear and nearly colourless, while the magnesian micas are dark reddish-brown or black, and the latter show strong dichroism, when the planes of section do not coincide with the basal planes of the crystals.

When schorl occurs in granites it may usually be recognised by the strong bluish tint which it here and there shows, when examined under the microscope by ordinary transmitted light, and also by the approximately triangular transverse sections which the crystals frequently exhibit.

When in thin sections of granite, magnetite and pyrites are present, they both appear opaque, and, to distinguish between them, it is necessary to examine them by reflected light, when their differences of colour and lustre become apparent.

With regard to the origin of granite, there has been considerable discussion, in which most antagonistic opinions have been brought forward; theories of its igneous, aqueous, and metamorphic origin having all been strongly advocated. Its eruptive character is inferred from the granite veins, which in certain localities traverse older rocks in a most irregular manner, while the dykes of quartz-porphyry which often emanate from, and can be traced to underlying granitic masses, and are, indeed, mere differentiations of granite, afford additional proofs of its eruptive character.

According to Hermann Credner, however, the mineral matter of the granitic veins in Saxony is not derived from deep sources, but from the partial decomposition of the adjacent rocks by the infiltration of water, and he observes that the mineral characters of the veins are influenced by those of the rocks which they traverse.[1]

With regard to the larger bosses and the huge granitic masses, from which such dykes and veins are given off, we can scarcely deny to the parent masses the origin which must be attributed to their offshoots, but, in the absence of such veins and dykes, it is easy to understand how, with considerable show of reason, a metamorphic origin may be assigned to those masses which, though once deep-seated, are now exposed by the denudation of enormous thicknesses of once overlying rock, and the question rather naturally

---

[1] 'Die Granitischen Gänge des sächsischen Granulitgebirges,' Hermann Credner, *Zeitsch. d. deutsch. geol. Ges., Jahrg.* 1875, p. 218.

arises whether they have not resulted from the metamorphism of sedimentary deposits, once so far beneath the earth's surface that they lay within a zone of comparatively high temperature. Admitting this, it seems that we are admitting no more than the conditions, or phases of the conditions, under which all eruptive rocks have been formed, and which are, therefore, just as fully entitled to the appellation of metamorphic rocks. That the passage sometimes observed from granite into gneiss is a proof that granite is the extreme phase of the metamorphism of sedimentary rocks does not always appear to be conclusive, since instances are known in which foliation is not indicative of bedding, and a few cases are recorded in which gneiss actually occurs in veins. In the present conflicting state of opinion upon this subject it behoves examination candidates to accept and cite the different opinions commonly held and set forth in the various manuals of geology. The student may afterwards judge of their respective merits from his own observations.

One of the arguments against the igneous origin of granite is that in granite the quartz has a specific gravity of 2·6, identical with that of silica derived from aqueous solution, while the specific gravity of fused silica is only 2·2.

This observation, in conjunction with many others, appears to have influenced to some extent the deductions of Professor Haughton, in his annual address to the Geological Society of Dublin in 1862. After giving a table of the relative specific gravities of natural and artificially fused rocks, he concludes in the following words :—

'It appears to me that the column of differences' (in the specific gravities of natural and artificially-fused rocks) 'greatly strengthens the argument of those chemists and geologists who believe that water played a much more important part in the formation of granites and traps than it has done in the production of trachytes, basalts, and lavas, and that they owe their relatively high specific gravity to its agency.'

'The only manner in which it seems possible to reconcile

the opposite theories of the origin of granite, derived from physical and chemical arguments, is to admit for granite what may be called hydro-metamorphic origin, which is the converse of what is commonly called metamorphic action, but which might more properly be designated pyro-metamorphic action. The metamorphism of rocks might thus be assumed to be twofold. Hydro-metamorphism, by which rocks, originally fused, and when in liquid fusion, poured into veins and dykes in pre-existing rocks, are subsequently altered in specific gravity and arrangement of minerals, by the action of water acting at temperatures which, though still high, would be quite inadequate to fuse the rock; and pyro-metamorphism, by which rocks originally stratified by mechanical deposition from water, come to be subsequently acted on by heat, and so transformed into what are commonly called the metamorphic rocks.'

'Granite, it appears to me, although generally a hydrometamorphic rock, may occasionally be the result of pyrometamorphic action; and such appears to have been its origin in Donegal, in Norway, and, perhaps, in the chain of the Swiss Alps.'[1]

This may be a very just opinion, especially if Professor Haughton does not imply, in his pyro-metamorphic action, the total exclusion of water from any participation in the changes effected. The two conditions of metamorphism which he indicates, most likely represent; in the hydrometamorphism, the presence of a large proportion of water and a moderately high temperature; in the pyro-metamorphism, a comparatively small portion of water and a much higher temperature. Such at least is a probable construction to put upon these conclusions; but, so far as metamorphism in its vulgar acceptation is concerned, there seems no reason, apart from the distinctions just given, for regarding granite

---

[1] 'On the Origin of Granite,' an address delivered before the Geological Society of Dublin, by the Rev. Samuel Haughton, F.R.S. Dublin, 1862.

as a metamorphic rock any more than basalt or trachyte, which have, in a certain sense, resulted from the extreme alteration of other rocks. The crystalline schists, gneiss, &c., are but phases of the conversion of sediments into true eruptive rocks; and, if the degree of alteration be put out of the question, the crystalline schists, the plutonic rocks, and the volcanic rocks, all seem equally eligible for the term metamorphic. In questions of metamorphism, it appears that the *nature* of the change is the first thing to consider; its *cause*, the next; its *degree*, the last.[1]

The different varieties of granite and of granitoid rocks may be summed up under the following heads :—

Porphyritic granite, in which the felspar crystals are large and well developed, being frequently several inches in diameter, as in those of Cornwall, Dartmoor, Shap, &c.

Various grades of texture occur between these granites and those which are termed fine-grained. When of the latter character, they pass into *micaceous felstones*.

Felstone (eurite,[2] hälleflinta, petrosilex), consists of felsitic matter, (viz., an intimate granular-crystalline, micro-crystalline, or crypto-crystalline, admixture of orthoclase and quartz, in which crystalline granules of plagioclastic felspars not unfrequently occur.) In this felsitic base, which, typically, constitutes the matrix of all felstones, felspar crystals, commonly orthoclase, are often developed ; and, like those in the porphyritic granites, are frequently twinned on the Carlsbad type. Such rocks are termed *felspar porphyries*.

---

[1] The less the unqualified term metamorphism is used, the better; since it merely implies change, without specifying the nature or extent of the change or the conditions under which the change took place.

[2] The terms Felstone and Eurite are frequently used synonymously; but eurite is stated by some authors to be more easily fusible than orthoclase, while the *eurite sursilicde* of Cordier is more difficultly fusible. The name Eurite is due to d'Aubuisson. Kinahan's definition of eurite, as a basic felstone (*Handy-Book of Rock-Names*, p. 48), might lead the unwary to regard it as a rock containing less than 60 per cent. of silica, but he is probably, to some extent, right in keeping the distinction between eurite and felstone, although at times rocks of an intermediate character are met with.

If, in such a felsitic matrix quartz occurs porphyritically either in crystals, but more usually in roundish blebs, the rock is termed *quartz-porphyry*; but it is common to find porphyritic orthoclase crystals also developed in quartz-porphyries. It seems, however, probable that the groundmass of true quartz-porphyries should, in many cases, rather be regarded as a very fine-grained or micro-crystalline granite. Rocks of this class are called elvans by the Cornish miners, and, indeed, in that district, the term elvan is very loosely applied. As, however, the dyke-forming rocks of Cornwall are mostly offshoots from the granitic masses of that district, the term elvan has for the most part been applied to more or less fine-grained or porphyritic granitoid rocks, and it is now, as a rule, regarded as a synonym for quartz-porphyry, or, as some authors term it, quartz-felsite.

Granitite is a term given to those varieties of granite which contain a certain amount of plagioclase (oligoclase). The orthoclase, in the rock to which this name has been applied, is flesh-red, and this mineral and quartz are the two principal constituents. The mica is a blackish-green magnesian mica, but it is usually present only in small quantity.

Since plagioclastic felspars exist, though in a subordinate capacity, in many granites, it seems that no line of demarcation can be drawn between them and the granitites.

Cordierite-granite is a variety occurring in certain localities in Norway, Greenland, and Bavaria. It is characterised by containing cordierite or iolite; this mineral partially, and sometimes wholly, replacing the mica. A greenish oligoclase is often present in the rock.

Luxullianite is composed of schorl, flesh-coloured orthoclase and quartz. The schorl, which is black, or greenish-black, is distributed in irregular nests or patches, and contrasts strongly in colour with the other constituents of the rock. Boulders of this stone occur in the neighbourhood of Luxullian, in Cornwall; and the late Duke of Wellington's sarcophagus was made from one of them. The rock has not, however, been met with *in situ*.

In a paper by Professor Bonney, published in No. 7 of the 'Mineralogical Magazine,' 1877, two varieties of tourmaline are stated to occur in this rock, and some evidence is adduced to show that this mineral is a product of alteration.

Aplite or *Haplite* (from ἁπλόος, simple), also termed semi-granite (*Halb-granit*) or granitell, is a rock of limited occurrence, consisting of a crystalline-granular admixture of felspar and quartz. The so-called graphic-granite or pegmatite is a structural variety of this rock, in which the quartz is developed in such a manner that it roughly resembles Hebrew characters, a polished surface of the rock appearing closely inscribed, whence the name 'graphic.'

Granulite (Weiss-stein or leptinite) is also composed of felspar and quartz, the felspar being orthoclase or microcline. It is more or less finely crystalline-granular, and frequently has a foliated or schistose character. It generally contains garnets, which, under the microscope, appear as little irregular, roundish, singly refracting grains, like drops of gum. Mr. John Arthur Phillips has observed double refraction in the garnets of some granulites.

This rock, in its schistose structure and mode of occurrence, seems to bear much the same relation to felstone that gneiss bears to granite, and it may therefore be classed with the metamorphic rocks. It often contains schorl and hornblende microliths, and occasionally sphene. The variety called trap-granulite contains plagioclastic felspars, and is somewhat poorer in silica.

Greisen (*Zwitter*, *Stockwerks-porphyr*) is a granular-crystalline rock, consisting of quartz and mica, the latter usually lithia-mica. Quartz is, however, the predominating constituent. When orthoclase occurs in it the rock passes into granite. Tinstone (cassiterite) is very commonly met with in greisen, either in small strings and veins, or in little crystals or granules. It is a rock of common occurrence in Saxony and Cornwall.[1]

Gneiss.—This term, in its proper sense, signifies foliated granite; but foliated rocks, consisting to a very great extent of hornblende and quartz, have also been styled gneiss, although they should rather be termed schistose amphibolites. Indeed, the name seems to have been somewhat loosely applied to foliated crystalline rocks of variable mineral constitution.

True gneiss differs in no way from granite, except structurally. A foliated structure is its essential peculiarity. It is sometimes interbedded with other rocks, and frequently exhibits stratification, which is often but not invariably coincident with the foliation. Darwin has shown that, in the gneiss of the Andes, the planes of foliation coincide with planes of cleavage. Sir R. I. Murchison pointed out that the foliation in some of the Scotch Silurian rocks corresponds with the planes of bedding; and similar observations have been made in Anglesey, by Professor Henslow; while, according to Professor Ramsay, and a host of other observers, the coincidence of foliation with bedding is of extremely common occurrence. To crystalline rocks, which exhibit this structure, the adjective *gneissic* is applied, a good practice, when the rock deviates in mineral composition from a true granite.

Gneiss has been split up into numerous varieties, which, in the main, are identical in mineral constitution with the corresponding varieties of granite. Thus we have, in addition to ordinary gneiss, oligoclase gneiss, a foliated rock corresponding with oligoclase granite, dichroite gneiss, adularia gneiss, garnet gneiss, syenitic gneiss, &c., &c.

Protogine is a gneiss in which, in addition to the ordinary constituents of granite, a greenish, pearly, or silvery talcose

---

[1] A paper, 'On some of the Stockworks of Cornwall,' by Dr. C. Le Neve Foster, was read before the Geological Society, London, January 9, 1878.

mineral is present. The rock, when not foliated, is termed protogine granite.

Cornubianite (proteolite) is a compact granular-scaly condition of gneiss, which is met with, at times, at the contact of granites with slates.

The accessory mineral constituents which occur in gneiss are very numerous, and are similar to those which occur as accessories in granites.

Those rocks which in mineral constitution and in structure more or less resemble granites, are spoken of as granitoid rocks. They range from the coarsely crystalline to the micro-crystalline or crystalline-granular varieties.

Many of them have a felsitic matrix, their distinctive characters being due to the larger, or porphyritic, development of one or more of their mineral constituents.

The felsitic matrix of these rocks consists of an intimate micro-crystalline, or granular admixture of felspar (mostly orthoclase) and quartz.

The following table might be greatly extended so as to embrace all the chief rocks, both basic and highly silicated. Thus, for instance, if plagioclastic felspar were substituted for orthoclase, then syenite would become diorite. The student, by constructing such tables, may thus, as his knowledge increases, see how far the classification of rocks is useful, and how they gradually pass from one type to another.

Gneiss, granulite, and several other rocks, have been described in this place because they are closely related to granite in mineral constitution; but they should, perhaps, in most cases, rather be classed with the metamorphosed sedimentary rocks.

TABULAR VIEW OF SOME OF THE PRINCIPAL DEVIATIONS FROM THE GRANITIC TYPE.

In this table the letter F indicates felspar of one or more

species, but mostly orthoclase. M represents mica of one or more species. Q = quartz. H = hornblende.

Syenite	.	. F		H	cryst-granular.
Quartz-syenite	.	F	Q	H	cryst-gran.
Syenitic granite	.	F M	Q	H	cryst-gran.
Syenitic gneiss	.	F M	Q	H	foliated
GRANITE	.	F M	Q		cryst-gran.
GNEISS	. .	F M	Q		foliated.
Haplite	. .	F	Q		cryst-gran.
Granulite	. .	F	Q		schistose.
Quartz-porphyry		F	Q		Felsitic matrix—Q porphyritic.
Felspar-porphyry		F	Q		,, ,, F ,,
Felstone	. .	F	Q		,, ,,
Greisen	. .		M Q		cryst-gran.
Quartzite	. .		Q		gran. compact.

FELSTONE GROUP.

Felstone (eurite, petrosilex, hälleflinte, felsite).—Felstone is a more or less compact rock, those varieties termed hälleflinte and hornstone having a peculiarly flinty aspect, while, in other cases, the rock is either finely crystalline-granular or granular, sometimes porphyritic, often microporphyritic. In colour felstone varies very greatly—brick-red, brown, grey, yellowish, and greyish-white tints being the most common. Many varieties have a more or less conchoidal fracture, and all of them, before the blowpipe, are fusible on the edges of splinters to a white or speckled enamel. The eurites proper are more easily fusible than the felstones or *eurites sursilicées* of Cordier. In the compact and in the non-porphyritic examples no definite minerals can be detected with the naked eye or with a lens, and the same may be said of the matrix in which porphyritic crystals occur. Sometimes, but not commonly, they present an imperfect schistose structure, as in the varieties termed felsite schist. They differ considerably in chemical composition, the amount of silica which they contain varying from

Deviations from Granite as a Type. 215

PLATE III.

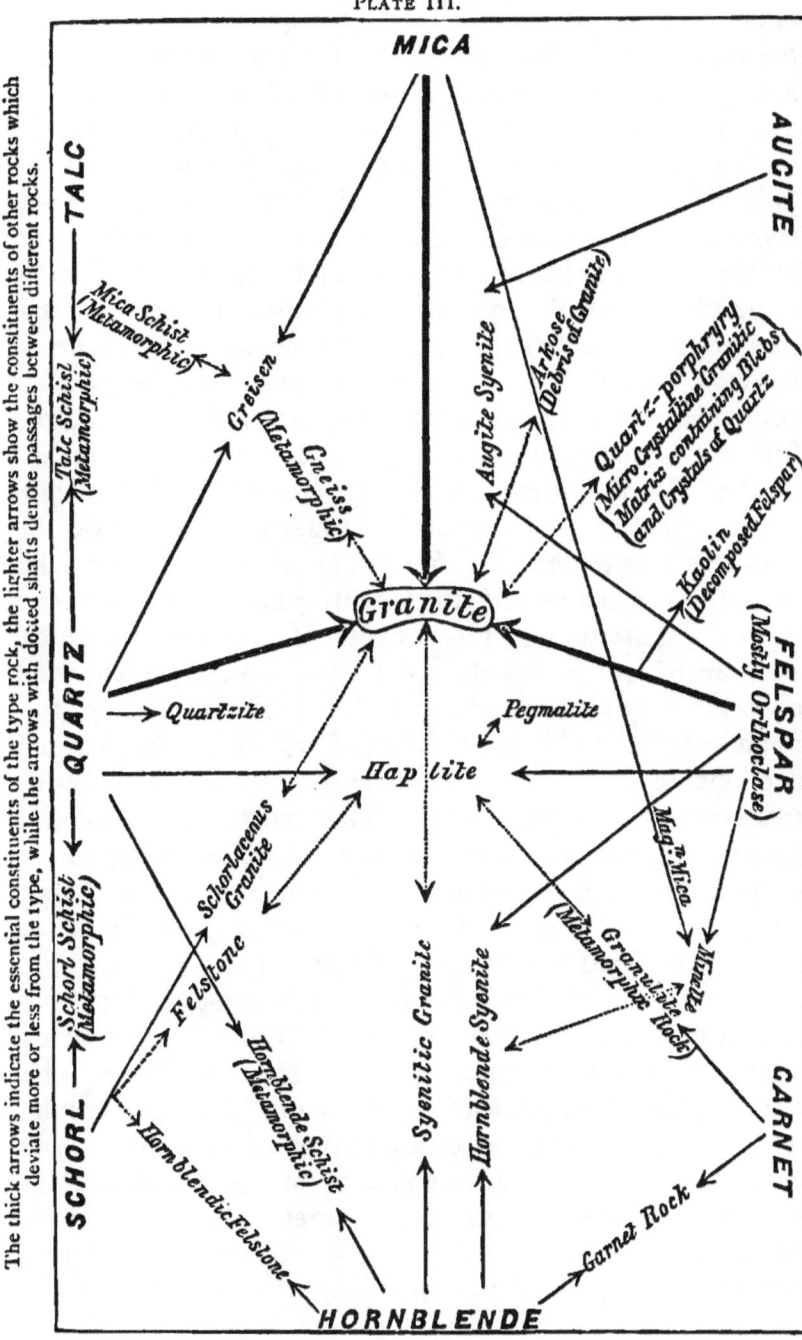

about 70 to 80 per cent., and they frequently have about 5 per cent. of the alkalies. Under the microscope they are also seen to vary greatly in character. Sometimes they show a micro-crystalline structure, in which, by polarised light, the section breaks up into a many-coloured mosaic, and the individual granules may be distinguished and identified, some of them as felspars, others as quartz, in others there is a somewhat similar but less defined structure, crypto-crystalline, in which individual minerals cannot be recognised. In some rare cases a considerable amount of true vitreous matter may be detected, lying between the micro-crystalline or granular component particles, or constituting the entire paste. More frequently the rock is wholly micro-crystalline or micro-felsitic. In the latter case between crossed Nicols the substance behaves as an amorphous mass. In this case the structure may be granular, fibrous, or microlitic, the granular and fibrous structure seldom presenting any definite character or individualisation of the constituent granules and fibres. Sections of such a rock do not however always present total obscurity between crossed Nicols, but transmit a feeble light, in an irregular and fickle manner, as regards its distribution. Sections of felstone occasionally present a radial-fibrous structure under the microscope; these constitute the varieties known as spherulitic felsite. Fluxion-structure is sometimes to be observed in felstones. It is probable, however, that many of the felsitic rocks which show this are more or less closely allied to the rhyolites. It is quite possible that in many cases the micro-crystalline or micro-granular structure of felstones simply represents the devitrification of an originally glassy magma, but, as remarked by A. von Lasaulx, the felsite pitchstones frequently fail to present the microscopic structure so characteristic of felstones. It is nevertheless far from uncommon to find small patches in sections of pitchstones and other vitreous rocks, in which devitrification has resulted in the production of structure, strikingly analogous, if not identical, with that of felstones.

Hornblende, micas, sometimes potash, sometimes magnesian, magnetite, titaniferous iron, &c., are met with in felstones Felspars are, also, often porphyritically developed, and the rock then becomes porphyritic felstone or felspar porphyry. Quartz also occurs porphyritically at times, either in crystals, or in roundish grains; the rock then becoming a quartz-porphyry or elvanite.

Indeed, passages may occur from felstone into granite, syenite, and various rhyolites. Felstone is generally more or less porphyritic, and occurs in dykes, veins and interbedded sheets.

Syenite Group.

Syenite, in the acceptation of the term, as first employed by Werner,[1] is a crystalline-granular rock, containing from 55 to 60 per cent. of silica, and consisting typically of orthoclase and hornblende. In mineral constitution, therefore, it approximates to some of the trachytes. Sometimes the felspar is microcline, and plagioclastic felspars are nearly always present. Sometimes augite or mica take the place of the hornblende, and occasionally the rock contains more or less sphene and quartz.

The syenites may therefore be divided into three groups, viz., hornblende syenite, augite syenite, and mica syenite. When quartz is present in any notable quantity the rock passes over to quartz syenite, and thence, when mica occurs, into syenitic granite.

Hornblende Syenite.—Orthoclase and hornblende are the chief constituents. Triclinic felspar is usually present in

[1] For many years it has been a common practice to apply the name syenite to syenitic or hornblendic granite. At times there has been considerable difference of opinion about the application of the name, which was first used by Pliny (*Syenites*) for the rock quarried at Syene in Egypt. The stone occurring at that locality is hornblendic granite. Hornblendic granite seems, therefore, to have a decided priority of claim to the name Syenite, but petrologists have found it convenient to restrict its application to quartzless rocks, such as those here described.

variable quantity; it may generally be referred to oligoclase, but the amount is as a rule comparatively small. The colour of the rock mostly depends upon the colour of the orthoclase, which varies considerably, red, brown, and white being the prevailing colours. The orthoclase crystals are frequently twinned on the Carlsbad type. The hornblende is usually greenish-black, and the crystals are generally, but not invariably, short; long bladed or acicular crystals sometimes occurring. The mica is a dark magnesian mica, commonly biotite. Epidote, magnetite, sphene, and pyrites frequently occur as accessories in this rock.

In structure the syenites as a rule greatly resemble granites, and they also occur in large eruptive masses, bosses, or veins. The gneissic syenite sometimes occurs in considerable beds, especially in the Laurentian series of Canada. The foliated rock of Cape Wrath in Sutherlandshire, Scotland, is rather amphibolite schist than gneiss, and some of the so-called gneiss of the Hebrides may also be referred to hornblendic schists.

Augite Syenite is composed of felspars, which, as a rule, are mostly orthoclastic, but the plagioclastic ones occasionally, though rarely, predominate. Augite is frequently plentiful, and sometimes a little hornblende occurs, which, as pointed out by A. von Lasaulx, is generally of a uralitic character, implying subsequent alteration of some of the pyroxenic constituents. Biotite, apatite, magnetite, and sphene are also of common occurrence as accessories in the composition of augite syenite. According to V. Lasaulx sphene is less plentiful in those varieties in which orthoclase is the predominant felspar. Analyses show that the augite syenites contain one or two per cent. less of silica than the hornblende syenites.

Mica Syenite is by no means a common rock, Calabria being almost the only district in which it is met with to any considerable extent. It occurs mostly in the form of veins or dykes. The rock consists of orthoclase, sometimes more or less plagioclastic felspar, biaxial magnesian mica, hornblende, occasionally some augite, which is often altered into pseudomorphs of chlorite or delessite, as in the minette of Seifersdorf in Saxony, while apatite, calcite, magnetite, and pyrites are also of common occurrence in these rocks; but, as a rule, sphene is never met with in mica syenite. The calcite and pyrites are products of secondary origin.

According to Rosenbusch,[1] Zirkel, and other petrologists, mica syenite and minette are intimately related if not identical. The former author also points out that some minettes are to be referred to the augite syenites.

Minette.—The matrix or paste of minette appears, under the microscope, as granular or granular-crystalline matter, in which microliths frequently occur; the latter according to A. von Lasaulx, who classes minette with the felstones, consist of felspar and mica, and the preponderance of evidence shows that quartz is an exceptional constituent of the rock. Under these circumstances the matrix can hardly be designated felsitic, and upon this ground hinges the question whether minette should be classed with the syenites or with the porphyritic felstones and granites. If the absence of free silica in the matrix be proved, it is evident that the affinities of minette are closer to syenite than to granite, but minette occurs in veins and dykes in both of these rocks, and dykes of it are also met with in sedimentary deposits of Silurian and Devonian age. It seems in many cases that micaceous felstones approximate rather closely to minette. Kengott appears to entertain some such idea in his 'Elemente der Petrographie.' V. Cotta's definition of minette is 'a felsitic matrix containing much mica and sometimes distinct crystals of orthoclase or hornblende.' The true difficulty seems to lie in the imperfect knowledge which we as yet possess of what a felsitic matrix really is. If quartz be excluded from such a matrix, and it is generally stated that minette seldom contains that mineral, then

[1] *Mik. Phys. d. Mass. Gest.*, p. 122.

minette is a mica-syenite with a micro-granular or micro-crystalline matrix; if, on the other hand, quartz be present, minette may be closely allied to the felstones, micaceous felstone forming a transitional link. It is possible that both conditions occur, and, if so, it may become necessary to classify the minettes. In some of the mica-traps of the English Lake District the author has found both orthoclastic and plagioclastic felspars which, in addition to the magnesian mica, occur in well-marked crystals. In such cases the rock appears to hold a position intermediate between minette and kersantite.

If minette represent a condition of the syenites which are rich in orthoclase, then kersantite is allied to those which are rich in plagioclastic felspars, and, in such instances, it may be questioned whether the affinities of kersantite are not more in the direction of diorite, especially of the micaceous varieties of that rock.

Speaking approximately, minette is a rock which contains magnesian-mica and sometimes hornblende crystals in a micro-granular or micro-crystalline matrix in which felspar crystals, mostly orthoclase, are porphyritically developed, while kersantite is a somewhat similar rock in which the felspathic components are mainly plagioclastic.

Both minette and kersantite occur, as a rule, in dykes. They are included under the old term Mica Trap.

B. von Cotta in his remarks on syenite says: 'Properly speaking there are no varieties of composition to adduce, unless we consider as such those transitions into granite and diorite which are occasioned by the occurrence of mica, quartz, and oligoclase.'[1]

This seems a very just generalisation, implying a sharp definition of syenite. If the term be allowed the wide scope which some petrologists accord to it, we might as well term mica-basalt a mica-diorite. That instances may be found of rocks, which, in mineral constitution, form connecting links

[1] *Rocks Classified and Described.* B. von Cotta. Eng. trans. 1866, p. 179.

between very many, if not all, of the eruptive rocks there is no doubt, but sharp, or moderately sharp, definitions constitute the basis of all classification, and, if these be abandoned, petrological nomenclature, as it exists at present, becomes almost worthless.

A good account of the minettes, kersantite, and kersanton, by Delesse, will be found in vol. x. of the 'Annales des Mines,' for 1857.

TRACHYTE GROUP.

Trachyte.—The rocks which have been included under this name are exceedingly numerous, and the term in its present acceptation has still a very wide range. The usual constituents of trachyte are sanidine, oligoclase, hornblende, sometimes augite, magnesian mica, magnetite, titaniferous-iron, tridymite, and at times some other minerals, such as sodalite, hauyne, nosean, sphene, mellilite, leucite, and olivine, which may, for the most part, be regarded as accessories. Plagioclastic felspar is generally associated to a greater or less extent with the sanidine in these rocks; hence Rosenbusch,[1] Zirkel,[2] and Von Lasaulx[3] consider that their division into sanidine trachytes and sanidine-oligoclase trachytes is of little or no account. The first author suggests that they may eventually be classified by determinations of the presence or absence of tridymite, and he thinks it probable that, by noting the relative occurrence of hornblende, augite, and magnesian mica, the trachytes may be arranged in a series homotaxial with that into which the syenites have been divided. The sodalite-, hauyne-, and nosean-bearing trachytes appear to some extent to be analogous to the phonolites.

The more highly-silicated trachytes are comprised in the group of rhyolites, and, in part at least, constitute the rhyolites proper, whose vitreous condition is met with in obsidian, &c., as already pointed out. The name trachyte is derived

---

[1] *Mik. Phys. d. Massigen Gesteine,* 1877, p. 200.
[2] *Mik. Besch. d. Min. u. Gest.* 1873, p. 382.
[3] *Elem. d. Petrographie,* 1875, p. 278.

from τραχὺς (rough), in allusion to the rough, scraping sensation which the surfaces of these rocks usually convey when rubbed with the fingers.

Geologically, the trachytes have been divided into trachytes and trachytic lavas, but the characters, even microscopic, of the one, have not been found to differ from those of the other. The trachytes proper are mostly of tertiary or post-tertiary age. Some rhyolites are coeval with them, while others have a great geological antiquity; but, as yet, comparatively little is known of these old rhyolites.

The trachytes may be conveniently classified in the following manner:—

Rhyolites proper or liparites[1]	i. Quartz-trachytes ii. Sanidine-trachytes	allied to perlite and obsidian.
Trachytes proper	iii. Quartzless-trachytes and domites	allied to syenite and the quartzless porphyries, such as porphyritic felstone, &c.

Analyses of these rocks show the following approximate variations of the amount of silica which they respectively contain.

Quartz-trachyte or quartz-rhyolite   75 to 77 per cent. silica.
Sanidine-trachyte or sanidine-rhyolite   74 to 78   „   „
Quartzless trachyte or trachyte proper   62 to 64   „   „

From this it will be seen that, although in some of the sanidine trachytes little or no quartz can be recognised, even microscopically, yet they contain a considerably higher percentage of silica than the trachytes proper.

It will be well to begin, in each case, with a brief statement of the microscopic characters of the ground-mass, base, or matrix of each of these three types, since the megascopic appearance of these ground-masses affords little or no insight as to their mineral constitution or structural peculiarities;

---

[1] The term lithoidite has also been applied to these rocks.

while, without a tolerably precise knowledge of these characters, it is often difficult to discriminate correctly between the different types.

QUARTZ-TRACHYTE (Quartz-rhyolite, Liparite).

The matrix is generally micro-aphanitic, and contains moderate-sized grains of quartz. The red varieties contain more or less peroxide of iron, in a finely-divided state, or in thin films. A micro-crystalline-granular or micro-granitic condition is less common in the matrices of quartz-trachytes than in those of quartz-porphyries, but nevertheless it is often present. The matrix of quartz-trachytes appears, to the naked eye, as a compact, or very finely-granular, substance, often rough and porous; it sometimes resembles hornstone, or porcellanite, while, at others, it has a dull, earthy or kaolinised appearance. It varies considerably in colour, brick-red, reddish-grey and yellowish- and brownish-white being some of the most common. The principal bodies porphyritically developed in this matrix are crystals and crystalline granules of sanidine and quartz, while plagioclastic felspars, hornblende, and magnesian mica (biotite) are often also well developed.

The sanidine crystals in the quartz-trachytes very frequently show twinning on the Carlsbad type. These crystals often appear much fissured and fractured. Under the microscope they frequently exhibit a zoned structure indicative of successive stages of accretion, and they often show numerous inclosures of glass, gas and steam pores, well developed crystals of quartz, and microliths of various kinds. Plagioclastic felspars, although frequently present, occur, as a rule, only very sparsely in the quartz-rhyolites, and are often so altered that the characteristic twin lamellæ are scarcely, if at all, perceptible, since they are sometimes completely converted into kaolin.

The quartz occurs both in roundish grains and in definite crystals. These contain inclosures of glass, which are often bounded by planes, corresponding to those of di-hexahedral crystals of quartz. Fluid lacunæ are not yet known to occur in the quartz of quartz-rhyolites, except in two or three instances.[1]

This general absence of fluid lacunæ distinguishes the quartz of these rocks from that of granite in which fluid inclosures are so common.

A little biotite frequently occurs in the quartz-trachytes. Hornblende is seldom plentiful. Tridymite and garnets are occasionally met with, but neither of these minerals are common accessories. Magnetite is generally present, but only in small quantity. Hard, vesicular varieties of quartz-trachyte occur in some localities, and are known by the name millstone-porphyry. The vesicles are often lined with chalcedony or quartz. Nodules or balls of chalcedony and opal are met with in the Hungarian rocks. Some of the quartz-trachytes show a fissile, slaty, or slabby structure, which sometimes originates in the varying character of different bands which exist in the rock, or else in a parallel arrangement of the sanidine crystals.

SANIDINE-TRACHYTE (Sanidine-rhyolite).

The matrix of this rock is usually of an aphanitic or micro-crystalline character. Under the microscope it is seen to consist almost wholly of little felspar crystals, mostly orthoclastic, but among which plagioclastic felspars are seldom absent. The felspar crystals are usually interspersed either with glass or micro-felsitic matter. Occasionally, as in the sanidine-rhyolite of Berkum near Bonn on the Rhine, the ground-mass consists almost exclusively of minute sanidine crystals with microliths of hornblende, grains of glassy matter and magnetite. In this matrix no quartz is to be recognised, although the rock contains over 72 per cent. of silica. Sometimes the matrix of sanidine-rhyolites seems

---

[1] They have since been noted by Zirkel, from Nevada, U.S., and by the Author, from Llyn Padarn, N. Wales.

almost wholly amorphous, or shows a finely fibrous structure, often hazy in appearance, while it occasionally assumes a radiate arrangement around certain points, thus giving rise to spherulitic structure. The spherulites in a rock of this character at Tolcsva, near Tokay, attain from one to two inches diameter. Well-individualised quartz sometimes occurs in the matrix of sanidine-trachyte, sometimes none is visible.

The minerals which are porphyritically developed in these rocks are, for the most part, crystals of sanidine, either single, or twinned on the Carlsbad type, crystals of plagioclastic felspar, which, as a rule, show more decomposition than the sanidine crystals, magnesian mica, magnetite, and occasionally hornblende, tridymite, and sphene.

TRACHYTE PROPER (Quartzless trachyte, quartzless sanidine-porphyry, domite).

The matrix of true trachytes consists generally of an aggregate of colourless felspar microliths, which, by their arrangement in certain directions, frequently indicate fluxion. Spiculæ and granules of greenish hornblende and specks of magnetite are also, as a rule, plentifully mixed up with the felspar microliths. By rotation of the section on the stage of the microscope, a very small quantity of interstitial glass may usually be detected, by its persistent darkness between crossed Nicols. The general colour of the matrix of trachyte is very variable, but greyish, yellowish, and reddish-brown tints are the most common.

The larger porphyritic crystals which occur in trachytes are sanidine and sometimes plagioclastic felspars:[1] the latter are not, however, always present. Hornblende is common in these rocks, and magnesian mica also frequently occurs in small crystals or scales, which are visible to the

[1] The true sanidine-trachytes contain but very little plagioclastic felspars, and, in some instances, none.

naked eye. The sanidine crystals are usually traversed by numerous irregular fissures, along which they are often displaced or faulted, as though they had been subjected to strain and pressure. They are very commonly twinned on the Carlsbad type, and the same may be said of the very minute crystals of this mineral which occur so plentifully in the matrix.

The larger sanidine crystals are sometimes an inch or two in diameter, as in the well-known trachyte of the Drachenfels, in the Siebengebirge on the Rhine.

The triclinic felspars are, as a rule, developed only on a small scale; and, as observed by v. Lasaulx, their glassy and cracked appearance often renders it difficult to distinguish between them and the smaller sanidine crystals.

Small crystals and spiculæ of hornblende are common in many trachytes. To the naked eye they look black, or greenish-black; while, when seen in thin sections by transmitted light, they appear green or brown.

Magnetite and apatite are also present, as a rule, in considerable quantity, the former mineral frequently forming black, granular envelopes around the hornblende crystals.

Tridymite sometimes occurs as a constituent of the matrix, and also in small cavities and druses in the rock. Sphene, hauyne, sodalite, nepheline, and specular iron are not of uncommon occurrence, while augite is sometimes, but rarely, met with in trachytes. In some of the ejected trachytic blocks, as in those of the Laacher See, a great number of mineral species occur, including, besides those already mentioned, zircon, corundum, meionite, garnet, spinel, staurolite, nosean, olivine, leucite, and various zeolites.

The name *domite* (from the Puy de Dôme, in Auvergne) has been applied to trachytes which contain a high percentage of silica, in some instances over 68 per cent., considered to be due to the presence of tridymite, since quartz is never observed in these rocks, while the former mineral occurs rather plentifully in the granular-microlitic matrix, which sometimes contains a small quantity of vitreous matter.

Scheme of Deviations from Trachyte as a Type.

PLATE IV.

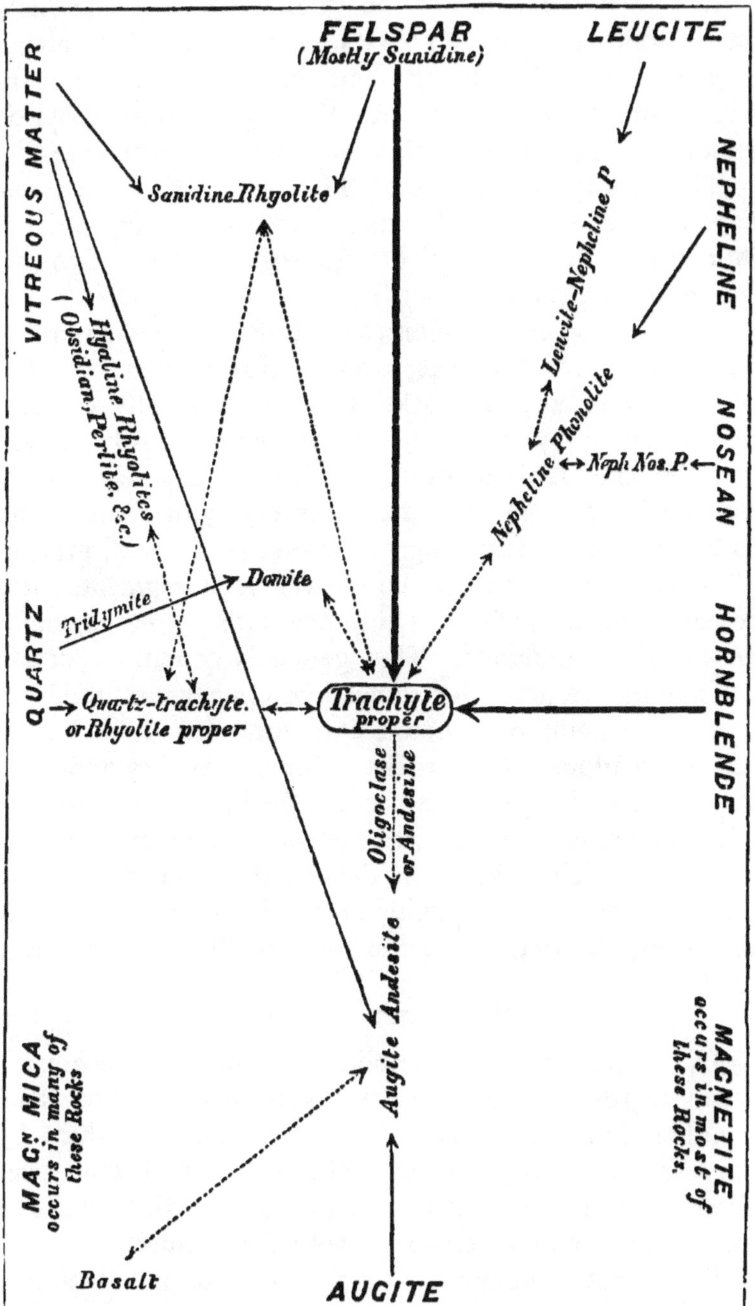

In mineral constitution the domites do not materially differ from ordinary trachytes, their somewhat higher percentage of silica being their chief characteristic, as pointed out by Zirkel and Rosenbusch. The porphyritic crystals in them, however, seldom attain any great size.

The domites form some of the most conspicuous dome-shaped hills or 'puys' which constitute such striking features in the scenery of Auvergne.—(Vide Scrope's 'Volcanos of Central France.')

Trachytic conglomerates and tuffs are composed of fragments of trachytic rocks, together with fragments of other eruptive and sedimentary rocks; these are frequently rounded, and cemented by crumbling earthy matter, mostly derived from the fine detritus resulting from the disintegration of trachytes. The tuffs are either of an earthy or granular and sandy character, mostly light-coloured—pale buff, grey, or yellowish-white. They contain fewer rock-fragments than the conglomerates, but pass into the latter as the fragments become more numerous. They generally contain crystals of sanidine, biotite, and other mineral components of trachyte. It seems difficult or impossible to draw any hard line in the classification of these rocks. Sometimes they appear to have resulted simply from the weathering of trachytes, at others they have more the character of volcanic ejectamenta, ashes, &c., which have been deposited in water; and, in both cases, there is frequently an admixture, to a greater or less extent, of detrital matter derived from sedimentary rocks.

PHONOLITE GROUP.

The rocks termed phonolite or clinkstone are in a certain degree related to the trachytes proper. The name phonolite, from φονος, sound, was first given to them by Klaproth. Both this and the other name, clinkstone, bear reference to the ringing or clinking sound which slabs or thin fragments emit when struck with the hammer.

The constituent minerals of phonolite are sanidine,

nepheline, and generally more or less hornblende and magnetite. Nosean and hauyne are often present in considerable quantity, also leucite, and sometimes tridymite. The minerals which are less common, and less important as constituents of phonolite, are augite, olivine, sphene, zircon, apatite, and titaniferous iron. Oligoclase as well as sanidine occurs in some of these rocks, and these oligoclase-sanidine phonolites are so closely related to the trachytes that they have received the name of trachy-phonolite.[1]

The matrix or ground-mass of phonolite is microcrystalline, and presents either a rough and porous, or a compact, character. The colour is usually grey, of a yellowish or slightly greenish tint. It is partly soluble in hydrochloric acid, the soluble portion being represented by nepheline and zeolitic decomposition products of that mineral, while the felspathic portion of the matrix constitutes the insoluble part.

The smaller the percentage of the insoluble matter, the higher, as a rule, is the percentage of water which the rock contains, and this is usually accompanied by an increase in its specific gravity. The larger the percentage of silica which a phonolite contains, the less, as a rule, is its percentage of soluble material. The amount of silica in phonolites generally ranges from 50 to a little over 60 per cent.

The phonolites fuse easily, before the blowpipe, to a whitish or greenish glass, and yield more or less water when heated.

The phonolites may be divided, according to their dominant mineral constituents, into the following subgroups :—

Nepheline-phonolite, Hauyne-phonolite, Nosean-phonolite, and
Felspar-phonolite.

---

[1] A very complete account of the different varieties of phonolite is given in Bořicky's *Petrographische Studien an den Phonolithgesteinen Böhmens (Archiv d. Naturw. Landesdurchforschung v. Böhmen.* Band iii. Geol. Abth.) Prag, 1873.

The phonolites have, however, also been classified by Dr. Emanuel Bořicky in the following manner :—

Nepheline-phonolite.
- i. *Nepheline-phonolite.*—With a compact matrix, and containing much nepheline, and porphyritic crystals of sanidine.
- ii. *Leucite-nepheline-phonolite.*—With a matrix of leucite and nepheline, much pyroxene, amphibole, and magnetite, but with sanidine poorly represented.

Nosean- or Hauyne-phonolite.
- iii. *Nepheline-nosean-phonolite* (Nepheline-hauyne-phonolite).—Containing much nosean and some hauyne, with a little sanidine, pyroxene, amphibole, titaniferous-iron, and sphene.
- iv. *Leucite-nosean-phonolite* (Leucite-hauyne-phonolite).—Consisting mainly of leucite, together with some nosean or hauyne, and more or less nepheline and sanidine. The leucite occurs both of microscopic and megascopic dimensions.
- v. *Sanidine-nosean-phonolite* (Sanidine-hauyne-phonolite).—A light-coloured rock, speckled with nosean and with a variable amount of porphyritic sanidine.

Sanidine-phonolite.
- vi. *Nepheline-sanidine-phonolite.*—A greenish, yellowish-grey, or dark grey, slaty or compact rock, weathering greyish-white, containing numerous porphyritic crystals of sanidine, and a few of augite or hornblende.
- vii. *Oligoclase-sanidine-phonolite* (Trachy-phonolite).—Containing from about 5 to 30 per cent. of triclinic felspar.
- viii. *Sanidine-phonolite.*—Sanidine is abundant. Nepheline and nosean occur in variable quantity up to 30 per cent., the former mineral constituting a large proportion of the matrix. Sanidine, augite, and hornblende occur porphyritically; also occasionally a little mica and sphene.

The relative proportions of the different minerals which constitute the matrix of phonolite vary considerably, the soluble portion varying in its relation to the insoluble portion from 15 to 55 per cent. of the rock, according to A. von Lasaulx.

The microscopic character of the matrix is generally micro-crystalline, and, in the phonolites of some localities, consists almost exclusively of superposed layers of small, well-defined crystals of nepheline and sanidine, with little, sparsely scattered crystals of hornblende and magnetite.

No amorphous or micro-aphanitic substance occurs in the matrix of phonolite; but its micro-crystalline nature is not always clearly perceptible, owing to the transparent and colourless character which it often exhibits.

The sanidine is very commonly twinned on the Carlsbad type, and is frequently more or less altered. Minute crystals of nepheline, nosean, and hornblende, and granules of magnetite, are sometimes seen lying within the crystals of sanidine, and often appear closely ranged along the margins of the sections of these crystals, which also, at times, contain inclosures of glass.

Plagioclastic felspars are only of exceptional occurrence in phonolites.

The nepheline crystals, although they occasionally attain moderate dimensions, are, as a rule, very minute, especially those which enter into the constitution of the matrix. They show sharply defined boundaries, and in some phonolites are very numerous; while in others, the mineral is so poorly represented, that the rocks approximate to trachytes. The nepheline crystals frequently show signs of alteration which, in its ultimate phase, results in the development of zeolitic matter, probably natrolite. Other zeolites, such as stilbite, thomsonite, chabasite, analcime, apophyllite, &c., also occur in phonolites.

The decomposition commences by the development of a yellowish fringe, which gradually passes from the exterior to the interior of the crystal, and, until these fringes unite, a nucleus of unaltered nepheline remains.

Haüyne is often very plentiful, and there are few

phonolites in which it is totally absent, except in those rocks in which leucite takes the place of nepheline. Hauyne and nosean are so closely related, both chemically and morphologically, that some mineralogists regard them as one species, and Rosenbusch includes them both under the older name hauyne. After treating a section containing these minerals with a drop of hydrochloric acid, it is possible to distinguish the sulphate-of-lime-bearing hauyne from the sulphate-of-soda-bearing nosean, by examining the section under the microscope, since, after a little time, the decomposition of the hauyne gives rise to little needles of gypsum, frequently associated, if a gentle heat be previously applied, with little rhombic, cube-like, doubly-refracting crystals of anhydrite.[1] These minerals vary considerably in colour, appearing brown, blue, yellow, green, black, and colourless. Some observers consider that, in their normal condition, they are colourless, and that, at all events, some of the colours are due to changes engendered by an elevated temperature, since colourless hauyne may be artificially coloured by heat. The decomposition of these minerals gives rise to the development of zeolitic matter, and also [when, as in the case of hauyne, they contain a fair amount of sulphate of lime] calcspar is formed.

Both hornblende and augite occur in some phonolites, and it is often very difficult to distinguish the one mineral from the other, since the augite in some cases exhibits strong dichroism, while in hornblende this character is sometimes quite absent. In such cases the angles of intersection of the cleavage planes, when they can be observed, afford a much safer means of discrimination than the phochroic characters of these minerals.

Magnetite is nearly always, and titaniferous-iron is occasionally, present in phonolites. Biotite, leucite, and tridymite are also often present in moderate quantity; while olivine, apatite, garnet, and zircon are among the less

[1] *Mik. Phys. d. Massigen Gesteine.* Rosenbusch, 1877, p. 218.

frequently occurring constituents. The phonolites which decompose most readily are, as a rule, those which are richest in nosean.

Phonolite occurs occasionally in the form of lava flows but more commonly in conical masses or hills. It sometimes exhibits well-marked columnar structure, and has a very general tendency to split into slabs or slates, the more finely-cleavable varieties being used for roofing purposes in certain localities. In advanced stages of weathering the rock passes into an earthy condition, known as phonolite-wacke.

Phonolite-conglomerate.—In some stages of disintegration phonolite-conglomerates are also formed; these consist of fragments of phonolite, and often of other rocks, together with fine, disintegrated phonolitic matter; the whole being frequently bound together by a calcareous cement. These conglomerates are mostly found at the bottoms of the phonolite hills, from which their materials have been derived.

Phonolite-tuff is an earthy rock of somewhat similar character, except that it contains but few actual rock-fragments. This earthy phonolitic matter often contains numerous crystals of the constituent minerals of phonolite, and the rock is generally cemented by more or less carbonate of lime. The eutaxites of the Canary Islands, and the piperno of Pianura, near Naples, are agglomeratic and banded lavas, which are considered to be more or less closely related to phonolite. The former have a partly vitreous character, and contain rock-fragments lying in tolerably regular layers, which impart a flecked or banded appearance to the lava, into which the fragments are partially fused. Want of space precludes any detailed account of these rocks, but descriptions will be found in the 'Geologische Beschreibung der Insel Teneriffe,' Fritsch u. Reiss. Winterthur, 1868; and in the works of Rosenbusch and v. Lasaulx, already cited.

ANDESITE GROUP.

The name andesite was first used by L. von Buch for certain rocks occurring in the Andes. The felspar in these rocks is plagioclastic, and is referred sometimes to andesine and sometimes to oligoclase. The other principal constituents are hornblende, augite and quartz, while more or less magnetite is also present as a rule. These rocks were first divided by Roth into hornblende-andesites and augite-andesites. The former are closely related to the trachytes, the latter to the basalts, and they thus constitute a connecting link between these highly basic and highly silicated rocks; a post also occupied to some extent, although upon different mineralogical grounds, by the trachy-dolerites.

In the rocks of both divisions of the andesite group quartz is sometimes present, sometimes absent, and, upon the presence or absence of this mineral, the andesites may be classed as

Hornblende-andesite { Quartzose hornblende-andesite or dacite.[1]
{ Quartzless hornblende-andesite.

Augite-andesite { Quartzless augite-andesite.
{ Quartzose augite-andesite (of doubtful authenticity).

Diallage- and hypersthene-andesites have also been described by Drasche.

Quartzose Dacite consists of a finely-granular or compact, grey, brownish or greenish-grey matrix, containing crystals of plagioclase (oligoclase or andesine) and sanidine, spiculæ of hornblende, and granules and crystals of quartz. Under the microscope, the matrix is seen to consist of microliths of plagioclase, sanidine and hornblende, together with fine grains of magnetite. Quartz seldom appears, according to A. von Lasaulx, to enter into the composition of the matrix when definite megascopic grains of quartz are visible in the rock.

As a rule, the matrix is entirely micro-crystalline, but at

[1] So named from its extensive occurrence in Dacia.

times, when examined microscopically, it shows here and there a very small quantity of interstitial glass.

The triclinic felspars are the most numerous and important of the porphyritic constituents of this rock, and analyses indicate that they may sometimes be referred not only to andesine and oligoclase, but also to labradorite. Crystals, showing interlamellation of triclinic felspar with sanidine, are sometimes to be seen under the microscope.

The quartz usually contains fluid lacunæ and magnetic dust.

The hornblende is either in spiculæ, or in well-developed little crystals, which sometimes show twinning, and seldom occurs in forms of purely microscopic dimensions. It shows strong dichroism, and often contains needles of apatite and grains of magnetite. Epidote and chlorite represent the ultimate phase of alteration of the hornblende. Occasionally crystals of augite may be detected in these hornblende-andesites. Olivine is never met with in them. Some of the hornblende-andesites of Hungary may be regarded as rhyolites, in which plagioclastic felspars play the part of sanidine.

The quartzose dacites have been divided into trachytic dacites, biotite dacites, &c.; in the latter hornblende is almost entirely absent, its place being represented by biotite. Some of these rocks are very poor in quartz, and they then pass into the quartzless hornblende-andesites.

The chemical composition of the dacites varies considerably in the amount of silica which is present, this fluctuation being due to the variable quantity of quartz which different dacites contain. Von Lasaulx gives as a mean analysis :— $SiO_2=66\cdot10$. $Al_2O_3=14\cdot80$. $FeO=6\cdot30$. $CaO=5\cdot30$. $MgO=2\cdot40$. K_2O and $Na_2O=7\cdot70$. $H_2O=0\cdot50$. In unaltered samples of dacite the soda is always in excess of the potash.

Quartzless Hornblende-Andesite.—These rocks mainly differ from the preceding in containing no quartz and little or no sanidine. Biotite and magnetite are plentiful in them.

Augite sometimes occurs, also nepheline. Hauyne and olivine are very rarely met with.

Quartzless hornblende-andesites are well represented in the Auvergne, the Siebengebirge and other localities, where they have commonly been designated trachytes.

A. von Lasaulx gives the following as a mean analysis of these rocks—

$SiO_2 = 59\cdot75 \,.\, Al_2O_3 = 17\cdot25 \,.\, FeO$ and $Fe_2O_3 = 7\cdot57 \,.\, CaO = 6 \,.$
$MgO = 1\cdot30 \,.\, K_2O = 3\cdot10 \,.\, Na_2O = 4 \,.\, H_2O = 1.$

Augite-Andesites.—The rocks which come under this denomination are closely related to basalt in their mineralogical constitution. They have a compact or finely-crystalline matrix, containing a considerable quantity of glass. The constituent minerals are triclinic felspar (either oligoclase or andesine), augite, magnetite, and at times more or less sanidine and hornblende, while occasionally quartz may be present. These rocks have been divided into quartzless- and quartzose-augite-andesites, but the latter appear to be of very exceptional occurrence, and it is probable that they approximate more, in some instances, to the hornblende-andesites.

The quartzless-augite-andesites frequently exhibit very distinct fluxion structure. The matrix is usually of a brown-grey or blackish colour. The glass which enters into the constitution of the matrix is sometimes quite clear and often contains trichites, at other times the glass is devitrified by the development of granular structure. Within this matrix crystals of oligoclase or andesine, augite and magnetite, are developed, and crystals of sanidine and biotite also frequently occur. Olivine and nosean are but rarely met with in this rock.

The felspar and augite crystals frequently exhibit inclosures of glass. The amount of sanidine present is always subordinate to that of the plagioclase.

The silica in the quartzless augite-andesites is sometimes more, sometimes less than 60 per cent.; the potash is a little over or under 3 per cent.; and the soda varies from a little over 2 to about 4 per cent.

The augite-andesites are essentially lavas.

Propylite is a rock closely allied in mineral constitution to the hornblende-andesites. It occurs very extensively in the United States, and bears, geologically, close affinities to volcanic rocks. Propylites also occur in Transylvania and Hungary, where, as at Kapnik and Nagybanya, they contain rich metalliferous veins. Propylites are of early tertiary age. They are in fact the first, or oldest, eruptive rocks of that epoch. The opinions of Stache and v. Richthofen appear to differ concerning this rock. Zirkel adopts the views of the former author, and says, ' That petrographical differences exist between propylite and hornblende-andesite cannot be any longer doubted.'[1] He describes some of the typical propylites as consisting of felspars completely filled with hornblende material, while the larger hornblende crystals are entirely altered into vivid-yellow epidote, slightly tinged with pale-green, occurring either in confused, fascicular, radiating aggregates, or in small roundish grains. Apatite is also present. Zirkel regards these rocks as more closely allied to porphyritic diorites than to andesites. His explanations, however, still seem to fail in giving a sharp definition for these rocks. The propylites are both quartzless and quartzose, like the andesites.

The propylites are the equivalents of the greenstone-trachytes of v. Richthofen, while the grey-trachytes of that author are represented by the hornblende-andesites.

Stache applied the term dacite to both of these groups. Zirkel objects to this extended application of dacite, and argues that propylite is an independent rock.

PORPHYRITE GROUP.

The rocks comprised under the term porphyrite are characterised by an aphanitic or micro-crystalline matrix, essentially composed either of triclinic felspar and horn-

[1] *Microscopic Petrology.* U. S. Exploration of 40th Parallel, p. 10.

blende, or of triclinic felspar and augite, in which larger porphyritic crystals of the same minerals are developed, while, in some instances, biotite takes the place of hornblende. From the mineral constitution of these rocks it is therefore evident that those consisting of triclinic felspar and hornblende are allied to diorite, while those containing triclinic felspar and augite may, since the felspar is usually oligoclase, be regarded as approximations to diabase. Those in which biotite acts as a substitute for hornblende are more or less closely related to the mica-diorites. The porphyrites are therefore divided into diorite and diabase-porphyrites.

These rocks may to some extent be considered as identical with diorite and diabase, differing from them chiefly in the fact that they have a micro-aphanitic, or in some cases a felsitic matrix, while diorite and diabase are crystalline granular throughout. In the diorite-porphyrites quartz is sometimes present, sometimes absent : in the diabase-porphyrites it is of very exceptional occurrence.[1]

Diorite-Porphyrites.—The matrix of these rocks is usually dark-brown or dark-grey, and may either be of a felsitic character, or it may consist of an admixture of microliths of oligoclase and hornbende. It is in both cases essentially crystalline, or micro-granular, never micro-aphanitic. Those rocks which possess a matrix of the former kind constitute the division of quartzose-diorite-porphyrites, and quartz occurs in them not only microscopically in the felsitic matrix, but also frequently in grains of moderate size which are visible to the naked eye. Those diorite-porphyrites which have a microlitic matrix and contain no quartz constitute the division of quartzless diorite-porphyrites. Occasionally, but rarely, the microscope shows the presence of a small quantity of vitreous matter in the matrix.

[1] It seems probable that in some instances a relation may exist between some of the porphyrites and eurites. Kinahan, in his *Handy-Book of Rock Names*, London, 1873, p. 49, ascribes a similar opinion to Naumann.

According to the nature of the minerals which are porphyritically developed in them, the diorite porphyrites may be distinguished as plagioclase-porphyrite, in which crystals of oligoclase are met with, while, at times, granules of quartz and crystals of biotite are also visible as porphyritic developments. Hornblende-porphyrite, in which distinct crystals of hornblende and triclinic felspar occur porphyritically. In some hornblende-porphyrites, in addition to oligoclase, a very little hornblende is occasionally visible. The mica-porphyrites are distinguished by the porphyritic development of biotite and oligoclase, the former mineral almost entirely supplanting the hornblende. Quartz is very generally present in this rock.

The rocks kersantite and kersanton which occur in the form of dykes in certain parts of Brittany are closely related to the mica-porphyrites. They consist of biotite porphyritically developed in a greenish-grey matrix, which consists in great part of oligoclase and which also at times occurs in well-developed little crystals. Kersantite differs mainly from kersanton in containing more or less hornblende. These rocks are sometimes amygdaloidal, and commonly contain minerals of secondary origin, such as chlorite, calc-spar, epidote, pyrites, &c.

Diabase-Porphyrites.—These rocks have a finely aphanitic or granular matrix, consisting of triclinic felspar and augite, and the same minerals also constitute the porphyritic crystals which occur in this matrix. The plagioclase crystals, which predominate over those of augite, are sometimes labradorite and sometimes oligoclase. Quartz is not of common occurrence in these rocks.

The diabase-porphyrites have been divided into diabase porphyry and augite porphyry: the term 'porphyrite' will, however, here be substituted for 'porphyry,' since the latter term implies the megascopic development of certain minerals, and, when coupled with the prenomen 'diabase,' may therefore be thought objectionable by those who consider that it is better in all cases to use the adjective 'porphyritic' and to abandon the term 'porphyry.'

Diabase-porphyrite, or plagioclase-diabasite, consists of crystals of augite and labradorite, or oligoclase, somewhat sparsely disseminated in a dark grey or greenish-grey matrix, which is compact and aphanitic, and consists of microliths and little crystals of triclinic felspars and augite, with occasional traces of vitreous matter. The felspars sometimes exhibit a greenish appearance, due to an admixture with chlorite. The pyroxenic constituents of the rock are frequently represented by pseudomorphs of serpentinous or chloritic substances (viridite). Calcspar and epidote are also among the usual secondary products. Magnetite occurs more or less plentifully. The verde-antique porphyry is one of the diabase-porphyrites.

Augite-porphyrite, or augite-diabasite, has also a compact aphanitic matrix in which porphyritic crystals of triclinic felspar and augite are developed, but the felspar is always labradorite, and the porphyritic crystals of augite are, as a rule, considerably in excess of the felspar crystals. The matrix, moreover, very frequently contains a certain amount of glass. Olivine crystals and pseudomorphs after olivine are also common in the augite-porphyrites.

Apatite, pyrites, chlorite, and calcspar are frequently met with, and occasionally a little orthoclase is present.

The augite is in some cases entirely, partially in others, converted into hornblendic matter, which is dichroic, and shows the characteristic hornblende cleavage. This product of the alteration of augite, which is known as uralite, is occasionally well developed in these rocks, which are thence termed uralite porphyries.

The augite-porphyrites occur in dykes or intrusive sheets, and in some instances they may represent lava flows. Sometimes they are vesicular and amygdaloidal. In the latter case the vesicles often contain many different species of zeolites and other minerals of secondary origin. Tufaceous conditions of these rocks occur in the Tyrol.

DIORITE GROUP.

The term 'greenstone,' which in its older signification embraced basalt, diabase, gabbro, diorite, &c., has subsequently been restricted in its application, and employed as a synonym for diorite. Since, however, the name greenstone is almost meaningless, it seems desirable either to discard it, or, still better, to use it in its original sense as an ambiguous and comprehensive term, useful in field geology, but otherwise only admissible as an expression of comparative ignorance, such as may safely be employed in the case of rocks of a certain type, which have reached so advanced a stage of decomposition, and in which the constituent minerals are so poorly developed, that it is no longer safe or possible to hazard any opinion concerning their precise normal mineralogical constitution.

Diorite is an essentially crystalline-granular admixture of triclinic felspar and hornblende. The majority of the diorites are quartzless, nevertheless quartz occurs in some of them, and they are then designated quartz-diorites. The triclinic felspar is sometimes oligoclase, sometimes labradorite, and this fact, again, gives rise to a division into oligoclase-diorites and labrador-diorites.

Oligoclase-diorite is a crystalline-granular admixture of oligoclase and hornblende. The texture of the rock varies from fine to coarse grained. The colour also is variable, being sometimes greenish-grey, at others greenish-black, while some of the coarser grained varieties have a speckled or blotched appearance. The rocks are sometimes very compact in texture, and are then styled diorite-aphanites, but in the most compact varieties it is seldom, even under the microscope, that any traces of a micro-aphanitic or of a devitrified paste can be detected. Glass inclosures and gaspores are not of common occurrence in the oligoclase crystals, but the latter sometimes contain fluid lacunæ. The oligoclase is usually white or greenish-white, and shows the twinning striation characteristic of triclinic felspars. Orthoclase is sometimes present in these diorites, but it plays a very subordinate part as compared with the oligoclase. The hornblende is mostly greenish-black, sometimes brownish, and occurs in long blade-like crystals or in imperfectly developed crystals, and irregularly shaped patches and grains. In the finer grained and aphanitic varieties of the rock the hornblende crystals are often exceedingly small. A microscopic examination of thin sections of diorite frequently shows the presence of numerous inclosures of glass, magnetite dust, and microliths, in the hornblende crystals, together with gas-pores.

Crystals of magnesian mica often occur in these rocks, and, when very numerous, they constitute the varieties known as mica-diorites. Apatite is present in nearly all diorites, and the little hexagonal prisms of this mineral may often, when sections are examined under the microscope, be seen to colonise in particular spots. Augite is occasionally, but not often, present. Chlorite, pyrites, magnetite, and titaniferous-iron are of very common occurrence in diorites, while garnet, sphene, and epidote are also met with as accessory constituents.

Labrador-diorite.—If labradorite be substituted for oligoclase, the foregoing description will answer equally well for this variety.

Quartz-diorite.—In the constitution of these rocks quartz plays a somewhat important part, and occurs both in megascopic and in microscopic, roundish or angular grains, but seldom in properly-developed crystals.

Fluid lacunæ are very numerous in some of these quartz grains, and, in addition to bubbles, they occasionally contain minute cubic crystals of rock-salt, as in the quartz-diorites of Quenast, in Belgium, described by Rénard. (See fig. 74, p. 165.) A very large number of diorites are quartziferous.

Scheme of Deviations from Diorite as a Type.

PLATE V.

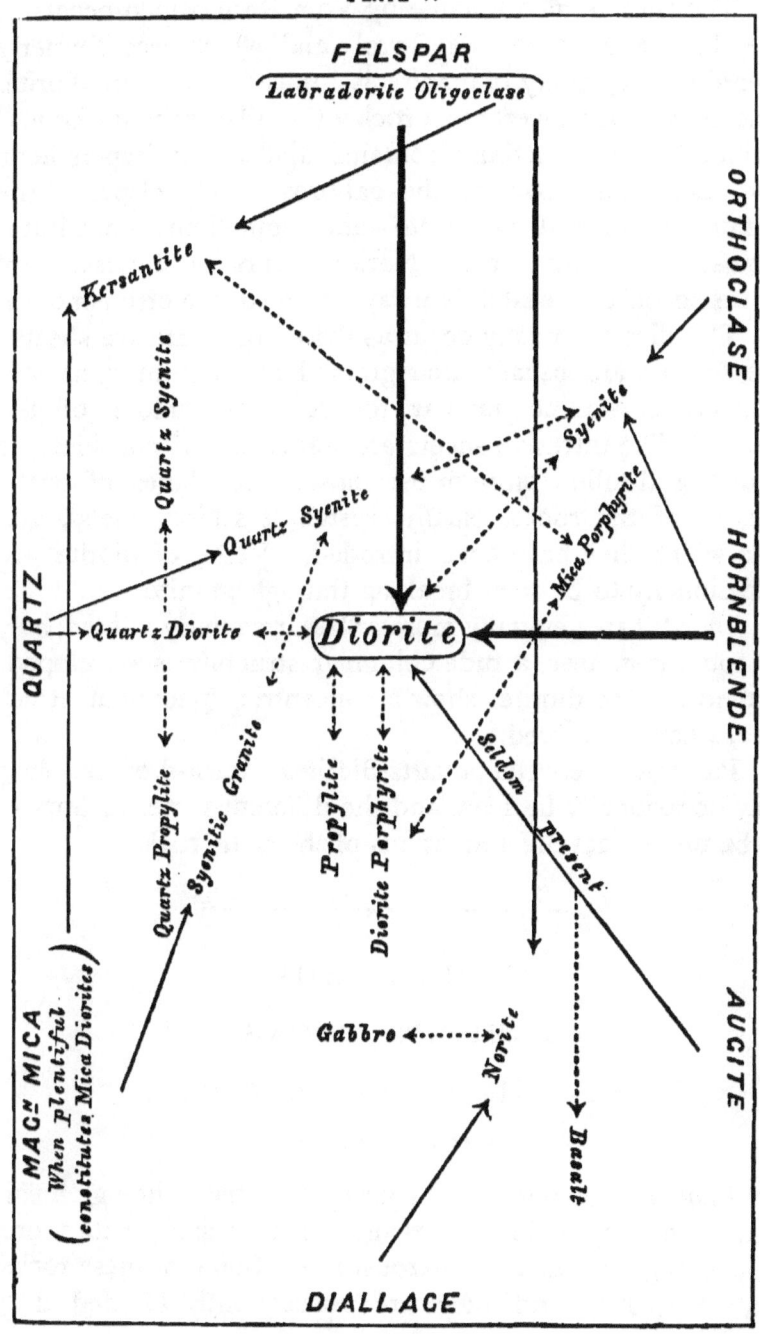

The rock termed tonalite by Vom Rath, which occurs in the Tonale Pass in the Tyrol, and which was formerly regarded as a variety of granite, is a micaceous quartz-diorite. The norite of Scheerer is a rock of similar mineral constitution, but it sometimes contains diallage or hypersthene and then passes over to the gabbros. The felspar is the dominant mineral in norite, and sometimes constitutes almost the entire rock. Norite occurs in the island of Hitteroe, off the coast of Norway, and also in North America.

The diorites mostly occur as dykes and intrusive sheets; the former are usually fine grained at their margins and coarsely crystalline-granular towards the middle of the dykes. The intrusive sheets are sometimes of considerable extent, and follow more or less closely the planes of stratification of the rocks, usually crystalline schists, gneiss, &c., into which they have been intruded. Veins of diorite are occasionally to be seen breaking through granite.

Diorites are generally more or less irregularly jointed, but, in some instances, a rude columnar structure is developed. Sometimes the diorites show a concentric spheroidal structure, when weathered.

Passages between quartz-diorites, mica-diorites, &c., may occasionally be seen, and the different varieties appear to be mere local differentiations of the same rock.

CHAPTER XIII.

ERUPTIVE ROCKS.

Class II.—Crystalline Rocks.

DIABASE GROUP.

DIABASE may be regarded typically as a crystalline-granular admixture of triclinic felspar and augite, usually with more or less magnetite and titaniferous-iron. Some of these rocks contain quartz, and they are consequently divided into quartz-diabase and quartzless-diabase, or diabase-proper. Chlorite has hitherto been cited as one of the essential constituents of the rock; but Rosenbusch very justly observes [1] that if the chloritic matter in diabase be, as it no doubt is, a secondary alteration-product, it then has only a 'pathological significance,' and cannot, therefore, rank as an essential constituent. That the viridite (chlorite, epichlorite, or chloritic matter) in diabase is a product of alteration is an opinion now very generally held, and, upon this ground, Allport has been led to regard diabase as an altered condition of dolerite. According to the researches of J. F. E. Dathe,[2] the felspar in diabase is not labradorite, but oligoclase. Rosenbusch considers that diabase aud gabbro are very closely allied, if not identical; and he bases this conclusion upon the argument that the essential difference between these rocks is represented by the statement that diabase contains augite, while gabbro contains diallage; and he argues that no essential difference, either chemical, morphological, or optical, exists between these two minerals, and that the only appreciable difference between them, which he is able to recognise, consists in the fact that diallage shows a structural condition of pinakoidal separation due to the presence of twin-lamellation, or interpositions, while, in augite, no such condition exists. The argument is most masterly; and the reader, if interested in this special question, should consult the original work.

Diabase (quartzless-diabase).—The rock is essentially crystalline in structure, and contains no trace of a glassy or of a devitrified base. The felspar is generally oligoclase, and, according to Dathe, remains unattacked when sections of the rock are treated with hot hydrochloric acid; but, in most cases, it has, microscopically, a hazy, granulated appear-

---

[1] *Mik. Phys. d. Massigen Gesteine.* Rosenbusch, Stuttgart, 1877, p. 323.
[2] *Mik. Untersuch. über Diabase.* J. F. E. Dathe. *Zeitsch. d. deutsch. Geol. Ges.* 1874. Bd. xxvi. pp. 1-40.

ance, due to partial decomposition. When this condition supervenes, the crystals seldom exhibit more than traces of their characteristic twin-lamellation; while, if the change be far advanced, all indications of this structure become obliterated, and the crystals may then at times be mistaken for orthoclase. In some instances the triclinic felspars, in microscopic sections of diabase, consist merely of two lamellæ, thus resembling Carlsbad twins of orthoclase; and occasionally, according to Rosenbusch, simple, untwinned crystals of triclinic felspar occur, which also simulate orthoclase. The presence of true orthoclase in diabase does not appear as yet to have been established. The augite crystals are generally seen, in microscopic sections, to be traversed by irregular fissures along which a green decomposition product (viridite), of a scaly, or occasionally a fibrous character, is developed; and this also occurs along the margins of the crystals. As this decomposition extends inwards, the crystals appear to be irregularly broken up into irregular patches, in the centres of which nuclei of unaltered augite still remain. When the decomposition is still further advanced, no traces of the original mineral are to be detected, and the viridite constitutes a complete pseudomorph after the augite. The plagioclase in these rocks seems also, at times, to undergo a somewhat similar alteration. Viridite also occurs interstitially between the different crystals, and the entire rock frequently appears to be thoroughly impregnated with this decomposition product. This green substance differs from chlorite by its more easy solubility in hydrochloric acid. It has been regarded by Gümbel as epichlorite, a mineral of intermediate character between chlorite and Schillerspar.

Apatite occurs plentifully in the coarser-grained diabases: in the more compact varieties it is less common. Biotite is also met with at times. Calcite is a very common secondary product in these rocks. Magnetite occurs in crystals and in fine grains. Titaniferous-iron, iron-pyrites, and copper-pyrites are also minerals of common occurrence in diabase, and in some few localities olivine forms one of the constituents. The percentage of silica in a diabase from the Harz is 44·6, while that in a quartz-diabase from Baden is 53·3.

Quartz-diabase.—The constituents of this rock are the same as those of quartzless-diabase, except that quartz and biotite are always present. Quartz-diabase is an essentially crystalline-granular rock, without any interstitial amorphous matter. It is, as a rule, coarse-grained in texture. The quartz occurs in small granules, seldom larger than a pin's head, and fluid lacunæ are plentiful in them. Olivine is occasionally to be seen in quartz-diabase. The diabases, both quartzless and quartzose, show considerable variation in their structural character. The following are some of the principal varieties :—

Granular-diabase, in which the individual constituents can be recognised with the naked eye.

Diabase-aphanite, a very fine-grained or compact variety, in which the constituents are not to be recognised without the aid of the lens or the microscope.

Calc-aphanite and *Calc-aphanite schist.*—Diabasic rocks which contain very numerous spherules of calcspar, bordered by chloritic matter, and appearing to pass into the surrounding matrix.

The calc-aphanite schist has, as its name implies, a schistose structure.

Diabase-schist is also an aphanitic rock with a schistose structure.

Amygdaloidal-diabase (*Diabasmandelstein*).—A vesicular diabase, in which the cavities have been filled with calcspar by infiltration. According to von Lasaulx, the character of the amygdaloids differs from that of the calcareous spherules in the calc-aphanites. To some of these rocks, occurring in Nassau and in the department of Haute Saone, in France, the name spilite has been applied.

Variolite is an aphanitic diabase of compact texture and greenish-grey colour, in which there occur little concretions of a paler colour, ranging up to the size of small nuts. The latter consist of concentric layers of plagioclase, augite, chlorite, and epidote. On weathered surfaces of the rock these little concretions form pustular markings, whence the name variolite. Von Lasaulx suggests that some of the rocks termed Schalstein may be referred to variolite.

Schalstein is, according to Gümbel, a tuff, or sedimentary deposit, the material of which has been derived from 'diabase-eruptions.' This cautious expression leaves it an open question whether the substance of these tuffs was derived from the disintegration of erupted rock, or whether it consists of fine ashy matter ejected from a crater. Certain schistose rocks, occurring in the neighbourhood of Brent Tor, appear closely to resemble some of the Nassau Schalstein. These schistose Devonshire rocks were regarded by Sir Henry De la Beche as volcanic ash. In most instances the evidence appears to show that they consist of eruptive matter of a diabasic character; but whether this finely divided matter was showered out as volcanic ashes, is not very evident; while the highly vesicular and amygdaloidal character which these beds sometimes assume renders it difficult to reconcile the co-existence of the vesicular with the schistose structure, unless the latter be regarded as superinduced.[1] The same doubt may be raised with regard to the amygdaloidal Schalstein of Nassau. In Gümbel's examination of the diabase tuffs, he mentions the appearance in them of a structure resembling the fluxion-structure seen in many lavas. This, however, he attributes to a totally different cause—namely, to the re-arrangement of detrital matter —and he distinguishes it by the term 'migration-texture.'

---

[1] 'The Eruptive Rocks of Brent Tor.' Frank Rutley, 1878, p. 36. The Author has since found that some of these rocks are true lavas in which a schistose structure has been developed.

J. Clifton Ward has pointed out the existence of a similar texture in some of the rocks in Cumberland.[1]

The schistose diabasic rocks contain a very large proportion of green chloritic matter, frequently in scales which are apparently often allied to sericite.

The greenstone-tuffs and diabase-tuffs are often closely allied to, or identical with, the rocks just described.

Gabbro Group.

The gabbros consist essentially of a triclinic felspar, generally labradorite, and diallage; sometimes, however, the felspar, when altered, is represented by Saussurite, and the diallage by hypersthene or smaragdite, and at times possibly by enstatite. Von Lasaulx classifies the gabbros in two groups, the gabbros proper and the hypersthenites; but in view of the researches of Descloizeaux on diallage, and of the opinions of Zirkel, Rosenbusch, and other petrologists, the hypersthenites, or those rocks which consist of rhombic pyroxene in conjunction with triclinic felspar, are of very restricted occurrence. Since, however, the rhombic minerals hypersthene and enstatite do occur in conjunction with plagioclase in a few rocks of limited occurrence, it seems desirable to follow the arrangement adopted by Rosenbusch, and to divide these rocks respectively into the plagioclase-diallage, or true gabbro; and the plagioclase-enstatite, or norite and hypersthenite sub-groups.

Plagioclase-Diallage Sub-Group.

Gabbro.—The structure of the gabbros is crystalline-granular or granitic, and no interstitial amorphous matter occurs in these rocks. Labradorite and diallage are the essential constituents. Olivine is sometimes present, and when this is the case the rock is distinguished by the term olivine-gabbro.

---

[1] 'Geology of the Northern Parts of the Lake District,' *Mem. Geol. Surv. Eng. and Wales.* J. C. Ward, 1877, p. 27.

The labradorite is usually pale-grey or white, and is easily fusible in the blowpipe flame. The crystals of labradorite frequently show signs of decomposition, and then contain green fibrous alteration products and opaque-white granules, and ultimately pass into Saussurite. The diallage occurs in tabular patches of a grey, brownish-green, or blackish-green colour, with a lamellar structure, seldom in distinctly-developed crystals. The orthodiagonal cleavage is well marked, and the cleavage-faces usually present a somewhat metallic lustre. The diallage is frequently fringed with a border of hornblende. The olivine-gabbros contain numerous dark-greenish grains of olivine, which are usually rich in microliths. They often show a fibrous, green alteration product, which probably represents their incipient conversion into serpentinous matter.

Augite, hornblende, biotite, talc, serpentine, calcspar, pyrites, pyrrhotine, garnet, and occasionally quartz sometimes occur as accessories in gabbro. The augite now and then forms interlamellations with the diallage, in the same way that interlamellations of monoclinic and triclinic felspars sometimes occur.

The three following analyses, cited by Von Lasaulx, show the general difference which exists chemically between ordinary gabbro, olivine gabbro, and the so-called hypersthenite of Penig, in Saxony:—

	i Gabbro (Bunsen)	ii Olivine-gabbro (Vom Rath)	iii Hypersthenite (?) (Bunsen)
SiO_2	= 51·35	50·08	49·90
Al_2O_3	= 19·82	15·36	16·04
FeO	= 14·95	6·72	—
Fe_2O_3	= —	—	7·81
MgO	= 4·14	9·99	10·08
CaO	= 3·51	14·90	14·48
K_2O	= 2·52	0·29	0·55
Na_2O	= 3·69	1·80	1·68
H_2O and loss	= 1· to 2·	1·27	1·46

A query has here been put against the rock of Penig, since, although for a long time regarded as a typical hypersthenite, it has since been suggested by Zirkel that the mineral in this rock, hitherto considered to be hypersthene, must now be reckoned as diallage, its almost total absence of dichroism precluding the supposition that it is hypersthene.[1] He also states that the supposed hypersthene in the so-called hypersthenites of Veltlin, Neurode, and the Isle of Skye, is simply diallage.

PLAGIOCLASE-ENSTATITE SUB-GROUP.

The rocks of this group appear to differ very little in mineralogical constitution from ordinary gabbros, except that their pyroxenic constituent is rhombic and not monoclinic. Considerable difficulty frequently attends the discrimination between hypersthene, enstatite, and bronzite, and it is therefore at times very unsafe to express any strong and decided opinion as to the precise nature of the rhombic mineral which represents the pyroxenic constituent of these rocks, which appear to be generally massed by Rosenbusch under the name norite. The rock of St. Paul's Island, on the coast of Labrador, in which the most typical hypersthene occurs, is placed by this author among the diallage- and olivine-bearing hypersthene norites. The norites of Hitteroe consist of plagioclase and hypersthene, in which the interposed plates, &c., so characteristic of the typical hypersthene, are very generally absent. These rocks also contain a little orthoclase and diallage. Olivine and mica occur in some of the norites, and bronzite has been recorded in one or two localities as an essential constituent.

Serpentine and Schillerspar (Bastite) are sometimes present in these rocks when they are more or less weathered. The norites never contain any glassy matter.

Gabbro occurs in the form of intrusive masses, often of

[1] *Mikroskop. Beschaff. d. Min. u. Gest.* Zirkel. Leipzig, 1873, p. 181.

considerable magnitude, and in dykes, veins, and intrusive sheets, which are sometimes forced along the planes of bedding in the adjacent stratified rocks.

BASALT GROUP.

Dolerite, anamesite, and basalt, or basaltite, are names applied to the rocks of this group, which imply different conditions of texture and crystalline development, rather than any marked difference in mineralogical constitution or chemical composition. Still some difference between them frequently exists in the relative percentages of silica which they contain, and also in their specific gravities.

The rocks of the basalt group all contain augite, magnetite, and titaniferous iron (of the last two minerals sometimes one, sometimes both are present), but they have in addition other mineral constituents which generally form a very considerable proportion of the rock, and indeed in some instances play quite a dominant part. Of these the felspars may claim the most prominent place. They are triclinic. Monoclinic felspar, although met with at times, is of comparatively exceptional occurrence. Olivine, nepheline, and leucite are minerals which exist very plentifully in some of the basalts; in the constitution of others they occupy quite a subordinate place; while in some, again, they are totally absent. The occurrence of leucite seems to be restricted to certain localities, and this mineral has not as yet been detected in any British rocks.

Hauyne and nosean (which latter may be included under the former name) are sometimes sparsely disseminated; at other times they occur in such considerable quantity as to give a distinctive character to the rock.

Micas occur rather plentifully in some of the basalts, occasionally to such an extent as to impart a special character to them.

The basalts have been conveniently classified by Möhl

according to their mineral constitution, in the following manner :—

   i. *Magma-basalts*, with a colourless or brown glass matrix.
   ii. *Plagioclase-basalts*, containing notably plagioclase and occasionally nepheline in addition to the essential augite, magnetite, &c. Leucite seldom.
   iii. *Nepheline-basalts*, containing notably nepheline, and sometimes leucite, in addition to augite, magnetite, &c. Plagioclase rare or absent.
   iv. *Leucite-basalts*.
   v. *Hauyne-* and *nosean-basalts*.
   vi. *Mica-basalts*.

The old term olivine-basalt is not included in this classification, apparently for the reason that olivine may, and very commonly does, occur to a greater or less extent in all of the basalts.

The rocks termed magma-basalts have already been alluded to under the name augite-tachylyte.[1]

The basalts vary considerably in structure : the coarsely crystalline varieties, and those in which the different mineral constituents are sufficiently well developed to be distinguished by the naked eye, are termed dolerites ; those in which the constituents are too small to be recognised without a magnifying power, but in which a crystalline texture is yet clearly discernible, are styled anamesites ; while the still more compact varieties, which, to unassisted vision, present a more or less homogeneous appearance, are called basalts (basalts proper) or basaltites.

Plagioclase basalts.—The constituents of these rocks are

---

[1] Bořicky classifies the basalts as
      Melaphyr-basalt,
      Felspar-basalt,
      Phonolite- and andesite-basalt,
      Trachy-basalt,
      Tachylyte-basalt.

Rosenbusch considers that most of the rocks included in Bořicky's last three groups are more or less closely allied to the tephrites, or those rocks which are characterised by the presence of nepheline or leucite plagioclastic felspars, augite, and magnetite. Titaniferous iron is frequently present. Apatite, olivine, nepheline, and hauyne may also be accessory: Carbonate of iron, calcspar, zeolites, chalcedony, &c., occur as secondary products, and very commonly fill the interior of vesicles. The spaces between the individual crystals are often filled with a glass-magma, usually of a brownish tint, and frequently containing great numbers of opaque trichites. As a rule, the glassy matter represents only a very small proportion of the entire rock. The plagioclase in these rocks is sometimes oligoclase, sometimes labradorite, anorthite, or andesine. It is, however, in most cases oligoclase. Orthoclase also occurs at times in these rocks, but its presence is quite exceptional. Olivine frequently forms an important constituent of the plagioclase basalts.

In microscopic sections of basalts which have undergone partial decomposition, the olivine and augite crystals are often merely represented by pseudomorphs of green matter, which is serpentine or some other hydrous silicate. The augite in basalts is generally rich in glass inclosures. Steam pores and fluid lacunæ are also of common occurrence in them. The olivine sometimes appears in tolerably well-defined crystals; but it is more usually in roundish grains, or in granular aggregates. The latter are sometimes of considerable size, and occasionally show, in external configuration, that they are large, rudely-developed crystals. The plagioclase basalts are of more frequent occurrence than any of the other rocks belonging to the basalt group.

in conjunction with plagioclase. Rosenbusch defines basalt as a rock consisting essentially of olivine, augite, and plagioclase, and regards these rocks as the tertiary and recent equivalents of olivine-diabase and melaphyre. Sandberger has proposed a division of these rocks into those which contain titanic-iron and those which contain magnetite. The former he designates dolerites, the latter basalts. This classification, however, as suggested by Rosenbusch, is by no means satisfactory, owing to the frequent difficulty in distinguishing between these two minerals, and also from the fact that magnetite is very commonly titaniferous.

Von Lasaulx gives the two following analyses as representing the average composition of a coarsely crystalline and of a compact variety, the former being a doleritic, the latter a basaltic, type:—

Plagioclase dolerite : $SiO_2 = 50.59$, $Al_2O_3 = 14.10$, $Fe_2O_3 = 16.02$, $CaO = 9.20$, $MgO = 5.09$, $K_2O = 1.05$, $Na_2O = 2.19$, $H_2O = 1.78$.

Plagioclase basaltite : $SiO_2 = 43.0$, $Al_2O_3 = 14.0$, Fe_2O_3 and $FeO = 15.30$, $CaO = 12.10$, $MgO = 9.10$, $K_2O = 1.30$, $Na_2O = 3.87$, $H_2O = 1.30$.

The lavas of Etna appear for the most part to be plagioclase basalts, rich in olivine. The plagioclase crystals in these lavas contain great numbers of irregularly-shaped glass inclosures.

Nepheline basalt, or nephelinite.—This is a crystalline granular admixture of nepheline, augite, and magnetite. More or less olivine is always present. Apatite, sphene, hauyne, melilite, and garnet are among the more common accessory minerals. The nephelinite of Katzenbuckel in the Odenwald, described by Rosenbusch,[1] may be taken as one of the most typical examples. Only mere traces of interstitial glass are ever to be seen in these rocks; some however contain interstitial nepheline, which may be easily distinguished from glassy matter by its polarisation, and by the crystalline aggregate character of the patches, although no definitely developed crystals may be visible.

The following is an analysis of the nephelinite of Katzenbuckel by Rosenbusch :—

$SiO_2 = 42.3$, Al_2O_3 and $Fe_2O_3 = 28.0$, CaO and $MgO = 13.65$, $K_2O = 2.73$, $Na_2O = 5.18$, $H_2O = 3.59$.

The rock also contains 0.65 per cent. of phosphoric acid, and traces of the oxides of nickel, cobalt, and manganese.

---

[1] *Der Nephelinit vom Katzenbuckel.* Freiburg, 1869.

Leucite, sodalite, and sanidine are occasionally met with as accessories in nepheline basalts. These rocks sometimes assume a very vesicular character, as in the millstone-lavas of the Eifel, and of Niedermendig on the Rhine. The latter rock is, however, often so rich in hauyne that it may rather be classed in the sub-group of hauyne-basalts.

These vesicular rocks assume an amygdaloidal character when the vesicles are filled with various minerals.

Leucite-basalt (Leucitophyr, Leucilite).—The rocks of this sub-group are seldom coarse-grained, and are mostly of a greyish colour, the leucite crystals often giving them a light speckled appearance. They are essentially aggregates of leucite, augite, and magnetite. Olivine and nepheline are very generally present, sometimes in considerable quantity. Nosean is sometimes plentiful, and biotite and sphene also occur as accessories. Under the microscope, scarcely any trace of vitreous matter is ever to be detected in the leucite-basalts, unless the leucite-sanidine-lavas of Vesuvius may be included under this name. In most of these rocks felspars are totally absent, although, in some of the leucitophyrs of Vesuvius and the Eifel, sanidine crystals are met with of tolerably large dimensions. In all the rocks of this sub-group, leucite is, as a rule, the dominant constituent. In some of the leucitophyrs, as for example in the rock termed sperone, which occurs in the neighbourhood of Rome, the leucite constitutes almost the entire mass, and the crystals, which are mostly of minute size, are very closely packed together. These crystals, when very small, no longer exhibit their characteristic crystalline form, but appear under the microscope as round spots having rather ill-defined boundaries. The leucite crystals are generally rich in interpositions, such as those previously described at page 110.

The leucite-sanidine lavas of Vesuvius have, as a rule, such a very complex mineralogical constitution, that they cannot be regarded as the equivalents of basalts. They number among their constituents leucite, sanidine, plagioclastic felspar [mainly anorthite], nepheline, sodalite, hauyne, augite, hornblende, olivine, biotite, apatite, &c. The majority of the Vesuvian lavas consist of seven or eight of these minerals. An account of them will be found in the Transactions of the Royal Irish Academy, vol. xxvi.[1]

Hauÿne-basalt (Hauynophyr).—Leucite, nepheline, hauyne, augite, and magnetite are the principal constituents, with usually some olivine and apatite. Vitreous matter occurs sparingly in these rocks and generally contains numerous trichites. Felspars, both monoclinic and triclinic, are absent. The hauyne crystals, which for the most part are blue, but also greyish or colourless at times, although frequently small, are seldom of very minute dimensions. Sometimes the rock assumes a porphyritic character, through the increased development of hauyne and augite. The most typical examples of hauyne-basalt occur at the Laacher See in the Eifel, and at Melfi near Naples. In the rock at the latter locality the hauyne crystals sometimes appear red, owing to the interposition of lamellæ of hematite. This red colour does not, however, always extend to the surface, so that the fractured crystals sometimes have a red nucleus surrounded by a blue border. Hauyne-basalts are rocks of very limited occurrence.

Mica-basalts.—These can scarcely be regarded as a distinct sub-group, since the mica which they contain does not exclude the occurrence, and cannot be considered as the representative, of any of the essential constituents of the sub-groups already described, unless, in any cases, its mode of occurrence could be reconciled with the observations of Kjerulf on the mica-pseudomorphs after augite, which he procured from the Eifel; or those of J. D. Dana, on the alteration of olivine crystals. The mica-basalts are rocks pertaining to the plagioclase, the nepheline, the leucite, or

---

[1] 'Report on the Chemical, Mineralogical, and Microscopical Characters of the Lavas of Vesuvius from 1631 to 1868,' by Professors Haughton and Hull. Dublin, 1876.

the hauyne basalts, and since any or all of these rocks may at times contain mica as an accessory, the only distinction which exists between them and the mica basalts appears to be summed up in the statement that mica basalts are *rich* in mica, while the other basalts contain that mineral in very limited quantity, or *as an accessory*. The mica crystals in these rocks vary considerably in size; sometimes they are quite large, at others, they occur as fine microscopic scales, distributed very closely and uniformly through the rock. These micas are mostly dark brown, reddish brown, or black, and may, in many cases, be referred to biotite.

The basalts occur in lava streams, plugs, intrusive sheets ('Whin Sill' of the north of England), and dykes. They are often traversed by structural planes which are, in some cases, so disposed that the rock assumes a columnar character, as at the Giant's Causeway, Fingal's Cave, and at many foreign localities. The columns are occasionally curved. They sometimes stand in vertical, at others in horizontal or inclined positions, which, in all cases, are directed at right angles to the surfaces upon which the rock cooled. This columnar structure is caused by the contraction of the basalt on cooling, but it is not exclusively in basalts that it occurs; it is occasionally to be met with in trachytes, phonolites, pitchstones, felstones, also in argillaceous rocks at their contact with eruptive masses.[1] Sometimes a platy or tabular structure is developed in basalt, especially near the margins of intrusive plugs or dykes. Spheroidal structure also occurs in these rocks, and the spheroids or balls may be seen often closely packed between the divisional planes which constitute the boundaries of the columns. The

---

[1] Some interesting experiments were made by Mr. W. Chandler Roberts in connection with the artificial production of columnar structure, and he has kindly supplied the following note. 'A mixture of clay and sand, in the form of Windsor-brick, was heated to about 1020° C. and slowly cooled. The mass was found to have contracted by about 6 per cent. (cubical), and columnar structure was well developed in it.'

columns are sometimes divided by cup-like joints, so that one portion of the column is convex and fits into a concave surface on the adjacent part of the column. The number of sides, which basalt columns present, varies. Occasionally they have only three sides, at other times five, six, or eight, as shown in the accompanying figure 87.[1] The subjoined papers on these structures may be consulted with advantage.[2] Basalt occurs in the form of wide-spread lava flows, and coulées or streams, in dykes, in irregular bosses, and in plugs or pipes, which represent the filled-up flues or feeders, from which lava streams were once poured out.

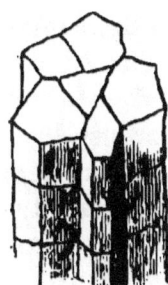
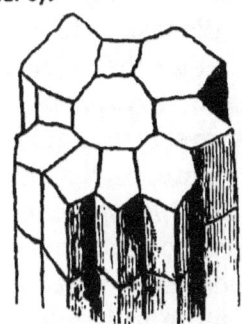

FIG. 87.

[1] From illustrations in the late G. V. Du Noyer's 'Notes on the Giant's Causeway,' *The Geologist*, vol. iii. 1860. 'It appears now to be pretty certainly established that the peculiar structure of columnar basalt is due to contraction and splitting, consequent upon cooling. The idea entertained by some of the older geologists, that the hexagonal form, so frequently found, was caused by the squeezing together of masses originally spherical, is geometrically incorrect. This process would give rise to rhombic dodecahedra, more or less regular, and could under no circumstances lead to six-sided columns. The cup-shaped joints, so frequently found, have also been shown to be a natural consequence of the contraction on cooling, to which the columnar structure is ascribed. In this view, the analogy of columnar basalt is rather to the splitting, often seen in the mud bottom of a dried-up pool, than to ordinary crystallisation. The direction of the columnar axis with reference to the apparent planes of cooling—the confusion of structure towards the middle of the dykes or beds—the cup joints—the irregularity of the prisms, whose cross sections are seldom regular hexagons—the way in which a hexagon passes into a pentagon through a heptagon, and not directly—all point to the contractile origin of the structure, at the same time that the result suggests a curious mimicry of imperfect crystallisation.'—C. W. M.

[2] Gregory Watt, 'Observations on Basalt,' *Phil. Trans.* 1804 pp. 279–313.

Scrope 'On Volcanoes.'

James Thomson, 'On the Jointed Prismatic Structure of the Giant's

Melaphyre.—The precise grounds upon which the rocks termed melaphyre have been raised to the dignity of a distinct petrological group are by no means apparent. Rosenbusch seems to regard them as closely related to, if not identical with, olivine-diabase. It is evidently a somewhat doubtful question whether they should be classed with diabase or basalt. Melaphyre may be defined as a fine grained or compact, black, greenish-black, or brownish-black aggregate of plagioclase, augite, olivine, magnetite, or titaniferous iron, and delessite or chlorophœite. These last two constituents are considered to distinguish melaphyre from basalt, [but melaphyres possess a vitreous, or a devitrified, magma which allies them more to basalt than to diabase].

Now, delessite is a ferruginous chlorite, and chlorophœite is a hydrous silicate of protoxide of iron, also allied to chlorite, or embraced by that very comprehensive term. Both of these minerals are decomposition products, and it therefore appears that their presence should serve to render the true nature of the rock a matter of doubt, rather than to constitute one of its distinctive characters.

Allport's suggestion that melaphyre should be included in the term dolerite, of which he regards it simply as a partially-altered condition, seems at least plausible.[1] The definition given by Bořicky in the introduction to his 'Petrographische Studien an den Melaphyr-Gesteinen Böhmens,'[2] appears to a great extent to confirm the foregoing statement. The melaphyres are of palæozoic age, and this

---

Causeway,' read at the British Association Meeting at Belfast in 1874, but only the title given in the report.

R. Mallet, 'On the Origin and Mechanism of Production . . . of Basalt,' *R. S. Proceedings*, 1874-5, vol. xxiii. pp. 180–84.

T. G. Bonney, 'On Columnar, Fissile, and Spheroidal Structure,' *Q. J. G. S.*, vol. xxxii. p. 140, 1876.

[1] 'On the Microscopic Structure and Composition of British Carboniferous Dolerites,' by S. Allport, *Quart. Journ. Geol. Soc.* vol. xxx. p. 530.

[2] *Archiv d. Nat. Wiss. Landesdurchforsch. v. Böhmen*, Geol. Abth. bd. iii.

Scheme of Deviations from Basalt as a Type.

PLATE VI.

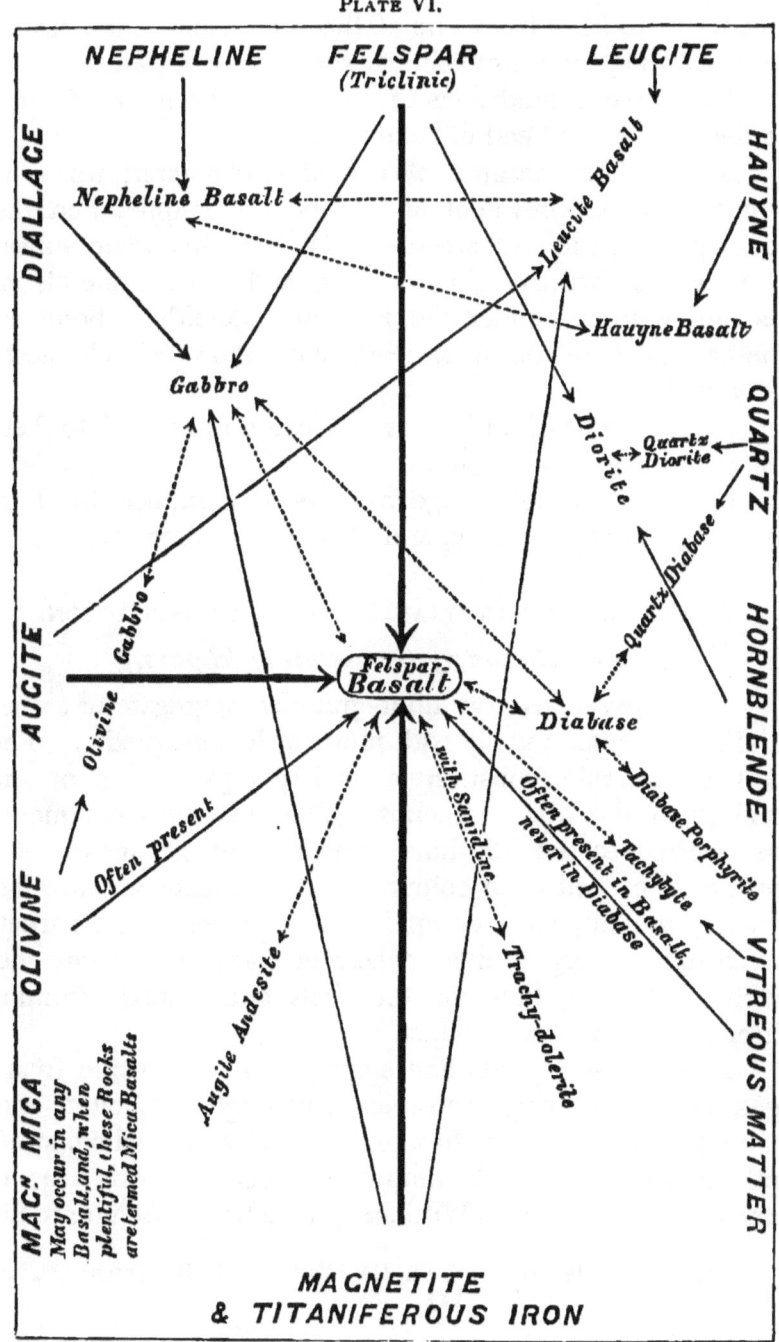

fact seems to have been one of the very insufficient reasons for separating them from similar rocks of later date.

The vitreous conditions of basalt have been already described under the head of tachylyte.

Basalts often assume a vesicular character, which is generally most prevalent at and near the upper and the lower parts of the lava streams. The vesicles, when subsequently filled with calcite, zeolites, and other minerals of secondary origin, render the rock amygdaloidal. Some of the basalts (toadstones) of Derbyshire show this character very well.

Basalts occur of various ages, ranging upwards into Tertiary, and Post-tertiary times.

Basalts of Dimetian age have been identified by Professors Judd and Bonney, and also by Mr. Tawney.[1]

ROCKS OF EXCEPTIONAL MINERAL CONSTITUTION.

(Characterised by the absence of felspars.)

Garnet-rock.—A crystalline-granular aggregate of garnet and hornblende, usually with more or less magnetite. The garnets, as a rule, constitute a far larger proportion of the rock than the other minerals. They may very commonly be referred to the iron-lime varieties, and are mostly of a brownish or yellowish colour. Other minerals are also frequently present, such as epidote and calcite. The garnet-rocks are of very limited occurrence, and are chiefly met with in Saxony, Bohemia, the Urals, and Canada, forming irregular veins in mica-schist.

Kinzigite.—A crystalline aggregate of spessartine (manganese-garnet), magnesian-mica and oligoclase, often containing some iolite and fibrolite, the latter a monoclinic mineral, having a chemical composition identical with that of andalusite. It occurs at Wittichen, at the Kinzig, Schwarzwald.

---

[1] 'On the Older Rocks of St. David's,' by E. B. Tawney, *Proc. Bristol Nat. Soc.* vol. ii. p. 113.

Eulysite.—An aggregate of reddish-brown garnet, green augite, and a mineral which, in chemical composition, is allied to the iron-olivine, fayalite. The last-named mineral is the dominant constituent of the rock. Eulysite occurs in a very thick bed in the gneiss of Tunaberg in Sweden.

Eklogite (Disthene-rock).—A granular aggregate of red or reddish-brown garnet, smaragdite (a green variety of diallage which, according to Descloizeaux, has the cleavage and optical properties of amphibole), hornblende, or omphacite (a grass-green variety of pyroxene with two sets of cleavage, one more perfect than the other, intersecting at an angle of 115°). Kyanite (disthene), silvery white mica, quartz, olivine, zircon, apatite, sphene, oligoclase, and pyrites, also occur at times as accessories. The eklogite from Eppenreuth contains about 70 per cent. of omphacite and 25 of garnet. Other varieties, such as those from the Fichtelgebirge and Baden, are, on the other hand, particularly rich in hornblende. Others again contain a large proportion of disthene, mica, and quartz, and on this account may preferably be termed disthene-rock.

The garnets in eklogite are often surrounded by an envelope of bright-green hornblende, while brown and feebly-dichroic hornblende also occurs in the same rock.

The freshly broken surfaces of the rock present a very beautiful appearance from the juxtaposition of red garnets with bright green omphacite.

Lherzolite.—A granular or crystalline-granular aggregate of olivine, enstatite, diopside, and picotite (a black spinel, containing over 7 per cent. of sesquioxide of chromium). The olivine is the dominant constituent. The rock varies considerably in texture; in some instances it is coarsely granular and feebly coherent, crumbling when handled; in others it is of a medium crystalline-granular character, and quite tough. The enstatite is of a greenish-brown or yellow colour. In thin sections it appears almost colourless by ordinary transmitted light. It has a more or less fibrous aspect. The cleavages parallel to ∞ P intersect at an angle of about 87°; less distinct cleavages parallel to the pinakoids are also visible, and are generally rendered more apparent by rotating the section between crossed Nicols. The diopside has a rough or stepped appearance on the abraded surfaces of sections, and shows the characteristic cleavage of augite. It occurs in roundish, green grains. The picotite appears, under the microscope, in very irregular brown, or (according to Bonney)[1] deep olive-green, patches or grains, which, in aspect, somewhat resemble dots and streaks of some gummy substance. They appear dark between crossed Nicols.

The olivine is very frequently altered into serpentine, the process of decomposition taking place in the first instance along the cracks in the olivine grains and crystals; and, as it advances, they become traversed by a mesh-work of little strings of serpentinous matter, until, in the final stage, no olivine remains, the rock often being impregnated with this decomposition product to such an extent that it is virtually a serpentine rock, as pointed out by Von Lasaulx, and more fully described by Bonney, who states his belief that lherzolite is an intrusive rock.

Rosenbusch, in describing the extreme phases of alteration into serpentine, remarks that the pseudomorphs after the enstatite and olivine may be microscopically distinguished from one another by the rectangular, grating-like disposition of the fibrous structure in the serpentine, resulting from the alteration of enstatite or diallage, and the irregular character of the fibrous mesh-work which is set up in the decomposed olivine.

Lherzolite occurs in veins of limestone at the Etang de Lherz, in the Eastern Pyrenees, whence it takes its name. It is also met with in the Tyrol, the department of Haute-Loire, Nassau, Norway, &c. Pyrope occurs as an accessory

[1] 'The Lherzolite of the Ariége,' *Geol. Mag.* decade ii. vol. iv. p. 64.

in a serpentinous condition of this rock, in certain localities. The olivine bombs met with in some basalts are, according to Von Lasaulx, closely akin to lherzolite. The chemical composition of a Norwegian lherzolite is cited by that author as

$SiO_2 = 37\cdot42$, $Al_2O_3 = 0\cdot10$, $MgO = 48\cdot22$, $FeO = 8\cdot88$, $MnO = 0\cdot17$, $NiO = 0\cdot23$, $H_2O = 0\cdot71$.

Dunite (so named from Dun Mountain in New Zealand, which consists in great part of this rock and serpentine), is a crystalline-granular aggregate of olivine and chromic-iron; the former occurring in yellowish-green grains and the latter in black octahedra. Dunite passes by alteration into serpentine. The frequent association of chromic-iron with serpentine renders it probable that many serpentines may have resulted from the alteration of some rock analogous in mineral constitution to dunite. This rock also occurs in the south of Spain and in several other European localities.

Diallage and enstatite are present in small quantities in some varieties of dunite, which, under these circumstances, approximates to lherzolite.

Picrite is a blackish-green crystalline rock with a compact, black matrix, containing porphyritic crystals and grains of olivine. The matrix may consist of hornblende, diallage, or biotite, associated with magnetite and calcspar. The olivine constitutes nearly half the bulk of the rock. A small amount of vitreous matter containing microliths is sometimes present.

Schorl-rock, although previously mentioned in the description of the granitic rocks, may also be placed in this miscellaneous group. The constituents are schorl and quartz. Topaz, mica, and tinstone sometimes occur as accessories. It is intimately connected with granite, and, by the accession of orthoclase and mica, passes into the schorlaceous varieties of that rock.

VOLCANIC EJECTAMENTA.

These comprise dust, ashes, sand, lapilli, and volcanic bombs. They all consist of mineral matter which has undergone a variable amount of trituration and, which has been ejected, either in a solid condition, or in a state of fusion. The expulsion of this matter from craters is due to the explosions of steam and gases which occur within the volcanic vents. The lava, which is in a fused and viscid or pasty condition, naturally becomes injected more or less completely with steam and gases, the bubbles of which, when imprisoned in the molten masses and unable to escape, produce a vesicular or spongy texture; so that it is common in volcanic ejectamenta to find fragments of rock, varying in size from fine dust to large blocks, in which a cellular, or pumiceous structure exists. These vesicles are sometimes coarse, sometimes so fine that they are not discernible to the naked eye. Most of the fragments of rocks and crystals which are shot up from the crater, fall back again, unless there be a sufficiently strong wind blowing, to carry them away. The constant attrition against one another which they undergo during these repeated journeys, up into the air, and back again into the crater, tends to round off any angles which the fragments may possess; and the process, if repeated long enough, would reduce the whole to fine sand or dust. Violent explosions also affect the matter within the flue of the volcano, forming a large amount of finely-comminuted and dusty material, which is often carried by the wind for long distances, or, if projected in calm weather, falls in showers over the cone. In some of the high volcanic mountains in the Andes, the flow of lava streams over the snow and ice, which rests at high levels, occasionally causes inundations, carrying vast quantities of fine mud, termed moya,[1] composed of volcanic dust and ashes; and similar mud-in-

[1] Dr. Theodor Wolf, 'Der Cotopaxi und seine letzte Eruption am 26 Juni, 1877,' *Neues Jahrbuch für Min. u. Geol.* Jahrgang 1878, Heft ii. p. 167.

undations are also produced there by the bursting of subterranean reservoirs of water during earthquakes. 'Mud derived from this source descended, in 1797, from the sides of Tunguragua in Quito, and filled valleys a thousand feet wide to the depth of six hundred feet, damming up rivers and causing lakes. In these currents and lakes of moya, thousands of small fish are sometimes enveloped, which, according to Humboldt, have lived and multiplied in subterranean cavities.'[1]

Volcanic ashes commonly consist of small fragments of lavas, and crystals of felspars, augite, olivine, biotite, magnetite, &c., and, in general, there is a more or less close relation in the minerals which constitute volcanic ashes and sands, and the mineral constitution of the lavas which have been erupted from the same crater. Volcanic ashes very often contain particles, or fused drops, of vitreous matter, and the crystals which occur in ashes also frequently contain numerous glass inclosures. The plagioclase crystals which occur in the ashes of Etna are especially rich in glass inclosures, but the plagioclase in the Etna lavas also contains them in great quantity.[2]

Volcanic sand simply differs from ash in the constituent fragments being coarser. The *puzzolana* of Naples and the *gravier noir* of the Puy Gravenoire in Auvergne are volcanic sands, used in the manufacture of hydraulic mortar.

Lapilli are moderate-sized fragments of rock, usually scoriaceous lava, which have been ejected from a crater. They may either occur imbedded in deposits of ashes and sand, or they may, of themselves, constitute accumulations.

The ejected lapilli are sometimes pumice fragments and, at times, form entire volcanic cones, as in some of the craters in the Lipari Islands.

Volcanic bombs vary considerably in character, but, generally-speaking, they may be defined as masses of molten rock-

---

[1] Lyell's *Principles of Geology*, 9th edition, p. 348.
[2] *Etna, a History of the Mountain and its Eruptions*, by G. F. Rodwell, p. 138. London, Kegan Paul & Co. 1878.

matter, which, by rotation in the air, during their upward flight and downward fall, have assumed a more or less spherical form, and have wholly or partially solidified before again reaching the earth; in the latter case, the imperfectly-solidified mass sometimes becomes flattened, by impact on the surface upon which it falls. Such bodies are termed slag-cakes.

The identification of very old deposits of volcanic ash is not always an easy task. Where numerous lapilli of scoriaceous and other unquestionably eruptive rock occur in old indurated ashes, as in those of Brent Tor in Devonshire, it is comparatively easy to recognise the origin of the deposits; but when these fail, it becomes a matter of considerable difficulty to say with any certainty whether a rock formed of broken crystals, such as might characterise any lava, in conjunction with very finely divided matter, such as might be referred either to fine volcanic dust or to ordinary detrital sediment, really represents a volcanic ash, or is simply a sediment formed wholly, or partly, of the detritus of pre-existing eruptive rocks. Some of the rocks mapped as ash beds in the English Lake district have undergone a very great amount of alteration, so that their originally fragmentary character is only revealed by superficial weathering or by microscopic examination, and, when the alteration becomes extreme, it is hardly possible to distinguish them from compact porphyritic felsite. Some of these rocks closely resemble hälleflinta, and a determination of their precise origin is a difficult exercise for micro-petrologists. The recognition of ash deposits is sometimes rendered troublesome by an intimate admixture of ordinary sedimentary matter. Much yet remains to be done in the determination of old volcanic ejectamenta, a field of inquiry in which none but the most sceptic are likely to demonstrate the truth.[1]

[1] The student may advantageously consult the recent paper by Dr. Albrecht Penck, 'Studien über lockere vulkanische Auswürflinge.' *Zeitschr. d. Deutsch. Geol. Ges.* 1878.

Altered Eruptive Rocks.

The alterations which eruptive rocks undergo, subsequently to their formation, represent, in most instances, decomposition, often accompanied by pseudomorphous replacement of their constituent minerals, due to the chemical changes effected by infiltration of water, charged, either with carbonic acid, or carrying in solution various soluble mineral substances which it has taken up during its passage through other rocks. There are comparatively few, or no eruptive rocks which do not, to some extent, show traces of such alteration, and the pseudomorphs which they contain are so numerous and interesting that they may constitute quite a special branch of study. The admirable 'Recherches sur les Pseudomorphoses' by Délesse, indicate how much may be done, and yet remains to be done, in this field of inquiry. Some of the most characteristic pseudomorphs will be found mentioned in the descriptions of the various rocks in which they occur.

The following are a few rocks which have resulted from the decomposition of eruptive rocks.

Serpentine has, in some instances, been demonstrated as the result of the decomposition of such rocks as lherzolite, gabbro, &c. It is also quite possible that serpentine may sometimes represent the alteration of ordinary sedimentary rocks, especially magnesian limestones, as suggested by Jukes and other geologists, but good evidence seems as yet to be wanting upon this point. Serpentine is also stated to result from the decomposition of some gneissic rocks and other crystalline-schists, also from garnet-rock and eklogite.

Serpentine is a fine-grained, massive, compact, rather tough, but soft rock, of very variable colour, dark and light shades of green, greenish-grey, and deep red being the most prevalent. It is often very beautifully veined and mottled with other colours, and, where not much exposed to atmospheric influences, it forms a valuable stone for decorative work in architecture. It is easily turned in the lathe into columns or small ornamental articles, and takes a high polish. It often contains crystals of diallage, which, to some extent, add to the beauty of the stone. Serpentine is also frequently traversed by white veins of steatite, in which angular fragments of serpentine are sometimes imbedded.

Serpentine is essentially a hydrous silicate of magnesia. When pure it contains at least two-thirds of silicate of magnesia, but it is frequently impure, through admixture with silicate of protoxide of iron, sesquioxide of chromium, argillaceous matter, and carbonates of lime and magnesia. The variations in its colour are due to different states of oxidation of the ferruginous matter which the rock contains. Serpentine when heated yields on an average about 12 per cent. of water.

The minerals which occur as accessories in serpentine are very numerous, and are for the most part the same as those which are met with as accessories in the crystalline schists, with which rocks serpentine is very commonly associated and interbedded. To them, however, may be added chromic iron, picotite, bronzite, schiller-spar, hematite, both massive and as specular iron, dolomite, calcite, brucite, magnesite, hydrotalcite, native copper, copper pyrites, and copper glance. Gold and platinum occur in the serpentines of the Ural.

Serpentine seems especially to result from the decomposition of rocks which are rich in olivine. Professor Bonney states that the serpentine of the Lizard in Cornwall contains decomposed olivine, enstatite, and picotite, and, from the presence of these minerals, he regards the rock as altered lherzolite, similar to the serpentine into which he had previously observed the typical lherzolite of the Ariége to pass. He regards the Lizard serpentine as a truly eruptive rock, and considers that the sedimentary rocks which surround it had been metamorphosed before its intrusion.[1]

[1] 'On the Serpentine and Associated Rocks of the Lizard District,' T. G. Bonney, *Quart. Journ. Geol. Soc.* vol. xxxiii. p. 923.

Sections of serpentine, when examined by ordinary transmitted light, usually appear of a pale greenish or yellowish colour. By polarised light the substance commonly exhibits a more or less fibrous structure which displays very feeble polarisation, pale bluish-grey and neutral tints predominating. The crystals of olivine, when they are only partially altered, appear in disconnected fragments, with moderately strong chromatic polarisation, the spaces between the fragments being occupied by fibrous serpentine, which represents the incipient decomposition of the olivine along those irregular cracks by which the mineral is so frequently traversed, as in fig. 88, which shows part of a section of serpentine from Coverack Cove, in Cornwall, after a drawing by Professor Bonney. The speckled portion of the figure indicates unaltered olivine, the remainder serpentine. Serpentine frequently contains veins of a finely fibrous mineral, chrysotile, which may simply be regarded as a fibrous condition of the serpentine itself.

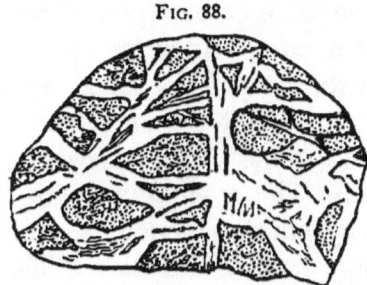
Fig. 88.

Occasionally an appearance of lamination, or fine bedding, is visible on the weathered surface of serpentine, the rock appearing to consist of thin alternating hard and soft bands, but the cause of this unequal weathering has not yet been satisfactorily determined.

Serpentine occurs either in intrusive bosses, in veins, or in beds interstratified with gneiss, mica-schist, chlorite-schist, talc-schist, &c.[1]

Laterite is a red, earthy rock, which occurs in beds lying between basalt and other lava flows, and results from their decomposition. It is strongly impregnated with sesquioxide of iron. Hematite and beauxite sometimes occur in beds of

[1] The views entertained by Dr. T. Sterry Hunt on the origin of serpentine will be found in his *Chemical and Geological Essays*.

this character. From the varying nature of the rocks from which it is derived, laterite has naturally a very variable composition, and indeed there is, as yet, no precise definition of this rock.

Palagonite-rock.—This results from the action of heated water or steam upon flows of lava, which effects the decomposition of many of the constituent minerals, and especially causes the peroxidation of any protoxide of iron compounds which the rock contains. The result is an amorphous, semi-vitreous substance of extremely variable colour (yellow, red, brown, and black). The chemical composition of palagonite corresponds more or less with that of the rock from which it is derived, except that no protoxide of iron remains, as it is all converted into sesquioxide, save in a few rare instances where magnetite occurs. The qualitative composition of the rock is represented by silica, alumina, sesquioxide of iron, magnesia, lime, soda, potash, and water. The percentage of silica mostly ranges between 30 and 40.

Under the microscope, palagonite appears as a perfectly amorphous substance, in which triclinic felspars, augite, olivine, undetermined microliths, and patches of colourless devitrified matter with a radiating fibrous structure occur. These last show dark interference-crosses in polarised light.

Palagonite-tuffs differ from palagonite rock in consisting not wholly of palagonite, but of fragments of that mineral, mixed with crystals of augite, olivine, and fragments of eruptive rocks.

Kaolin or China-clay is a soft, white, earthy rock, which results from the decomposition of the felspar in granites. When pure, it may be regarded as a bisilicate of alumina, plus two equivalents of water; but the composition varies. It may also result from the decomposition of leucite, beryl, &c., but all the important deposits of China-clay are in the main derived from orthoclase. These deposits are sometimes rather impure from the presence of other constituent minerals of granite. Some of them, in which quartz is plentiful, are termed China-stone.

The use of these clays for the manufacture of porcelain is too well known to need more than mention. The kaolin of Cornwall was first employed for this purpose by William Cookworthy of Plymouth in 1755. It has to be carefully levigated before it is fit for the potteries.

SEDIMENTARY ROCKS.

The general character of sedimentary rocks has already been described at page 15 *et seqq.* In this place, merely the lithological characters and industrial applications of the most typical varieties will be dealt with. These rocks constitute so large a proportion of the earth's crust, and have such an important bearing upon water-supply, agriculture, and mining and enginering operations, their application for building, road-making, and other industrial purposes is so extensive, and the history of their formation, and of the past conditions of life which existed at the time of their deposition, as shown by their fossils, presents so many points of scientific interest, that it would be impossible, even within the limits of a very large volume, to do even moderate justice to so great a subject. It has not been with any desire to underrate the importance of the sedimentary rocks that so comparatively large a proportion of this work has been devoted to the description of their eruptive brethren, but, because an elementary knowledge of their mineral constitution and structure is, when compared with that of the eruptive rocks, far more simple for the student to acquire.

The sedimentary rocks, as already stated, may be divided into two series, the unaltered or normal, and the altered or metamorphic. In the latter series, extreme phases of alteration carry the metamorphic rocks out of the sedimentary, and into the eruptive division. These eruptive rocks, when brought to the surface, and subsequently denuded, supply the materials from which fresh sediments are partly formed, so that petrology becomes the study of an endless cycle of changes from eruptive to sedimentary, and from sedimentary to eruptive rocks. The former class of changes are the result of atmospheric and marine denudation, and are due, more to mechanical than to chemical agency. The changes of the latter class are chemical and physical in their nature.

The altered or metamorphic rocks form, therefore, a transitional series between the unaltered, or normal-sedimentary, and the eruptive series. Sometimes the alteration is so slight that it is difficult to detect, and its precise nature still more difficult to demonstrate; at others, where great alteration has taken place, it is almost, and, in extreme phases, quite, impossible to say with certainty whether a rock should be referred to the metamorphic or to the eruptive series, since there is no natural boundary between them.

Before considering those which have been altered, it will be better to describe the

UNALTERED OR NORMAL SEDIMENTARY SERIES.

These rocks may be classed as arenaceous, argillaceous, and calcareous. They are, however, of a more or less mixed character as a rule, the arenaceous rocks being often cemented by calcareous matter, the argillaceous and calcareous rocks frequently containing a certain admixture of sand; while, again, some of the argillaceous series are impregnated with a variable amount of carbonate of lime, and those of the calcareous series are sometimes more or less argillaceous. Some of the normal sedimentary rocks contain fragments of felspars, scales of mica, and other detritus, derived from the disintegration of pre-existing eruptive rocks. The sedimentary rocks occur in strata or beds, which rest upon one another, which have a regular order of sequence, and which generally contain fossils of characteristic types. The remarks made in the earlier part of this work, and the numerous text-books of geology, which deal more or less fully with the stratigraphical and palæontological branches of the science, render it unnecessary to say anything here upon these subjects.

The materials of which sedimentary rocks consist are usually more or less rounded by attrition, the result of their transport by water, or, in the case of æolian rocks, of their transport by wind.

ARENACEOUS GROUP.
(*Sandstones.*)

These rocks consist essentially of grains of silica. They either occur as superficial accumulations of loose sand forming desert tracts, or low-lying districts on sea coasts, where the wind piles the sand up in dunes, or they may occur as beds of loose sand, interstratified with coherent beds of rock. They are also met with in a state of more or less imperfect consolidation, the grains being feebly held together by an iron-oxide or by calcareous matter; or they may be excessively hard and compact, the constituent grains being cemented by either silica, carbonate of lime, iron-oxides or carbonate of iron. The rocks called grits vary considerably in lithological character. The term 'grit' appears indeed to be very ill-defined. The millstone grit, which may be taken as one of the leading types, is more or less coarse-grained, while some of the Silurian rocks, such as the Coniston and Denbighshire grits, are frequently very fine-grained and compact in character. Under these circumstances it seems that a grit may best be defined as a strongly-coherent, well-cemented, or tough sandstone, usually, but not necessarily, of coarse texture. In some few cases there even appears to be, according to Prof. Morris, no cementing matter present, as in some of the new red sandstones, the constituent grains being apparently held together merely by surface cohesion superinduced by pressure. It is not possible within the limits of this work to do more than allude to some of the most important sandstones which occur in the British Isles. Those used for building-stone and paving are for the most part of old red, carboniferous, triassic, and neocomian age. Commencing with the oldest and lowest in the series, the *Cambrian* and *Silurian* grits are for the most part very tough, closely compacted sandstones, frequently containing minute fragments of felspars and sometimes scales of mica. Their constitution implies that they are formed, at all events to some extent, from the detritus of pre-existing eruptive rocks. They are, in some instances, fusible before the blowpipe, on the edges of thin splinters, which is probably due to their admixture with felspathic matter. They are generally traversed by numerous joints, so that they are seldom used for building purposes, except locally in the construction of rough walls. They are, however, well suited for road metal, and in some places good flagstones are quarried, but these are, for the most part, rather to be regarded as sandy shales and slates, than true sandstones. The flaggy sandstones are generally micaceous, and to this circumstance their fissile character is often due. Although some beds of sandstone and grit occur in the Devonian series, they are unimportant from an economic point of view, but in the *Old Red Sandstone* (the chronological equivalent of the Devonian series), both building-stones and flagstones are quarried. They are mainly employed in the districts where the stone is procured. It is often of a deep reddish-brown or purple colour owing to the presence of peroxide of iron; at other times it is greyish, occasionally with a greenish tinge. The stone, if judiciously laid, is tolerably durable, but in some old buildings, such as Chepstow Castle and Tintern Abbey, it has suffered considerably from the weather. Old red sandstone is extensively used for paving in many of the large towns in England, Scotland, and Ireland, and is also largely employed as a general building-stone and as road metal in the districts where it is quarried. The red sandstone quarries of Cork and Kerry yield in some instances building-stone of a very durable character.[1] The Dundee and Arbroath sandstones, known as Caithness flagstones, quarried on the east coast of Scotland, form good and durable material for paving and building, but the former is too sombre in colour to give a pleasing effect when used for architectural purposes.[2]

[1] *Building and Ornamental Stones*, E. Hull, p. 266. London, 1872.
[2] *Applications of Geology to the Arts and Manufactures*, Ansted, p. 153. 1865.

The *carboniferous* sandstones, including those of the Yoredale series, the millstone grit, and the coal-measures, are very important from an industrial point of view, since they afford good material for building and paving. The Halifax, Bradford, and Rochdale flags are extensively used for the latter purpose in the North of England, and are well known to builders under the name of Yorkshire flags. Some of them absorb water very readily, consequently, in very exposed and damp situations, they are liable to flake, especially if placed in positions where they are unable to part with their moisture. The stone from Bramley Fall near Leeds belongs to the millstone grit, and is largely used for architectural purposes. The Rotherham stone is worked for building purposes and for grindstones, and that at Hart Hill for scythe-stones. The Wickersley stone (Middle Coal-measure Sandstone) makes good grindstones.

The pennant grits and sandstones occurring in the coal-measures of the Bristol coal-field are important building-stones. The fine-grained pale-brown and grey sandstones from Craigleith, near Edinburgh, and the Binnie quarry in Linlithgowshire, are also extensively employed for buildings. They darken somewhat on exposure, but are amongst the most durable of building-stones. The Craigleith stone contains only about 1 per cent. of carbonate of lime, the cementing medium being mainly siliceous. A little mica and carbonaceous matter is also present to the extent of about 1 per cent., the remainder of the rock, 98 per cent., consisting of silica. According to Ansted, a cubic foot of Craigleith stone, weighing about 146 lbs., will absorb 4 pints of water, and good samples will resist crushing weights to the extent of 5,800 lbs. to the square inch. Of the carboniferous sandstones used in Ireland the Carlow flags are perhaps the most important: they are sometimes more or less micaceous, and are of dark bluish or grey colour.

The *Permian* sandstones, which are the equivalents of the Continental Rothliegende, or lower division of the Permian system, are but little used in this country, except locally, for building-stone, as in some parts of Cumberland,[1] Staffordshire, Nottinghamshire, and Yorkshire. At Mansfield, in Nottinghamshire, reddish brown and almost white varieties of triassic sandstone are quarried, and are said to be durable. As a rule the Permian sandstones are not well suited for building, being very absorbent and liable to decay. The Permian sandstone from the neighbourhood of St. Bees was used for the construction of Furness Abbey. These rocks have mostly a deep red colour, due to the presence of peroxide of iron, which, together with dolomitic matter, constitutes their cement.

Triassic sandstones.—Those belonging to the upper trias or Keuper are the most important as building-stones, the sandstones of the lower trias or Bunter being, as a rule, of too loosely-cemented and friable a character for such purposes. The latter are, however, used for moulds in foundries, and, occasionally, for buildings. They are often variegated and mottled, whence the name Bunter—from the German *bunt*, variegated or coloured—and they frequently exhibit false bedding. The Keuper series affords good building-stone, especially the lower Keuper sandstones, which are extensively used in the midland and north-western counties. It is of pale red, brown, and yellow colours, sometimes almost white, and is mostly fine-grained and easy to work. This stone has been largely used in the cathedrals of Chester and Worcester.

The loosely-coherent triassic sandstone of Alderley Edge, in Cheshire, is partly cemented by carbonates of copper and other mineral matter, derived from infiltrating solutions. To obtain the copper, the sandstone is crushed, the copper salts dissolved in sulphuric acid, and redeposited on scrap iron in the metallic condition. The sandstone yields but little more than 1 per cent. of copper. Keuper sandstone is quarried

---

[1] The Penrith and St. Bees sandstones are much used as building-stones.

in Antrim, and is stated by Professor Hull to be exceedingly well adapted for architectural purposes.

Amongst the carboniferous and triassic rocks of some countries a sandstone occurs to which the name Arkose is given. It consists essentially of the same constituents as granite, and has been derived from the disintegration of granitic rocks. Some valuable notes upon arkose occur in the 'Mémoire sur les Roches dites Plutoniennes de la Belgique et de l'Ardenne Française,' by MM. Ch. de la Vallée Poussin and A. Rénard, Brussels, 1876, p. 120.

Jurassic sandstones.—The rocks of the jurassic period are for the most part limestones, but good sandstone is quarried at Aislaby, near Whitby, in Yorkshire, and has been used in the construction of Whitby Abbey and several other important buildings.[1] In Lincolnshire, Northamptonshire, and Dorsetshire, sandstone, belonging to the inferior oolite, is employed for building. 'The ferruginous, or calcareous rock of the lower part of the Northampton sand is locally used for building purposes, but it does not usually possess much durability. The white sands in the upper part of the series are dug for making mortar.'[2]

Cretaceous sandstones.—Those of most importance, from an economic point of view, are the sandstones belonging to the Hastings sand series. The sand-rock of the Hastings Sands is not a very coherent stone when first dug, but it hardens on exposure, and, although locally used for building, it is not very durable. It is generally of a warm yellowish or brownish colour, and has a somewhat ferruginous cement.

Bargate stone is quarried at Godalming for building purposes. It is a calcareous sandstone, and occurs in the upper part of the Hythe beds. The rubbly sandstones in the Hythe beds in Kent are locally termed hassock. The Folkestone beds of the lower greensand afford hard sand-

---

[1] 'Mineral Statistics,' *Mem. Geol. Surv.* part ii. 1858. R. Hunt.
[2] 'The Geology of Rutland,' J. W. Judd. *Mem. Geol. Surv.* p. 92. 1875.

stone and grit, suitable for building and road-making. In the upper greensand at Godstone and Merstham, a pale calcareous sandstone called fire-stone occurs, which is well suited for the floors of furnaces, and is also a durable building-stone.[1]

Flints are procured from the upper chalk, and are extensively used as building material and road-metal.

Tertiary sandstones.—Although, in this country, beds of sand are of constant occurrence in the tertiary formations, they are not, as a rule, sufficiently coherent to be of value for building purposes, except for making mortar. They are, when pure, used in the manufacture of glass. The Headon Hill sands, which occur in the Bagshot series in the Isle of Wight, are largely used for this purpose. The less pure sands are applied to various other uses. There are, however, a few very hard tertiary sandstones, which are used in this country for building and paving, some of which are derived from the Woolwich series, and others from the Bagshot beds. In some parts of the world tertiary sandstones attain great importance. Sandstones of miocene age constitute a considerable part of the Himalayas. As an appendix to the rocks of this group we may place Tripoli, a fine white pulverulent deposit, which consists almost exclusively of the siliceous skeletons of diatoms.

The mud of the river Parret (which runs into the Bristol Channel) affords the material of which 'Bath bricks' are made. This mud has been stated to consist almost wholly of the remains of diatoms. Microscopic examination, however, disproves the statement, and shows that diatoms form but a very small proportion of the mud.[2]

---

[1] See also 'Geology of the Weald,' by W. Topley, *Mem. Geol. Surv. Eng. & Wales,* p. 371.

[2] Geology of East Somerset and Bristol Coal Fields,' *Memoirs of Geol. Surv.* p. 161. H. B. Woodward, 1876.

Argillaceous Group.

(Clays, Shales, and Slates.)

These rocks are, chemically speaking, impure hydrous silicates of alumina. Sometimes the impurity consists of sand, sometimes of carbonate of lime; and more or less carbonaceous matter is in many cases present. Their coarseness of texture is mainly dependent upon the coarseness of the sand which often occurs in them. When free from sand, they are usually of fine texture. They have all originally been deposited as mud, in most instances at the bottom of the sea, in others at the bottoms of lakes or as deltas, and, exceptionally, over land, when temporarily flooded by the overflow of rivers, as in the case of the Nile. Clay deposits often have a well-laminated structure, and, in the older geological formations, have assumed a more or less indurated character, frequently accompanied by a tendency to split along the planes of bedding. Very often another and more strongly-marked fissile structure is superinduced in directions cutting across the planes of stratification at various angles. This is slaty cleavage, described at page 35. Those argillaceous rocks which split parallel with the planes of lamination or bedding are called flags, but the term flag is applied to a rock of any character which splits along its bedding into large flat slabs, and consequently it is common to find the term used to denote sandstones which are sufficiently fissile, when quarried, to yield slabs or flags. To the argillaceous rocks which split in directions other than that of bedding the term slate is given. Still, in this case, the term is also applied to rocks which differ widely from ordinary slate. The Collyweston slates, calcareous sandstones of the inferior oolite, and the green-slates of the Lake District, which have been mapped as volcanic ash by the Geological Survey, are examples of the application of the term slate as indicative of fissile structure, and not of lithological character.

Cambrian slates.—These are very important rocks, affording compact roofing-slates of admirable quality, mostly of a dark purple or greenish colour, and capable of being split into very thin and large slates, exceedingly free from pyrites, which is common in many slates, but, from its decomposition, is most detrimental to them as roofing material. The slates of the Penrhyn and Bangor and of the Dinorwig or Llanberis quarries in North Wales are of Cambrian age.

Silurian slates and flags.—The Skiddaw [lower Silurian slates of Cumberland], are black, or dark-grey rocks, which are often traversed by many sets of cleavage planes, causing them to break up into splinters or dice, so that no good roofing-slate can, as a rule, be procured from them. The best lower Silurian slates of North Wales are quarried in the Llandeilo and Bala beds. They are black, dark grey, and pale grey. Ffestiniog, Llangollen, and Aberdovey are among the principal quarries. The cleavage in these rocks is often wonderfully perfect and even, so that occasionally slates ten feet long, six inches or a foot wide, and scarcely thicker than a stout piece of cardboard, are procured. These remarkably thin slates are tolerably flexible. The upper Silurian rocks also afford good slates and flags in certain localities, while the rough material serves for local building purposes. In many parts of the English Lake District the houses are commonly constructed of rough slates and flags derived from the Bannisdale and Coniston series. The quoins of the better class of these houses are often built of a light-coloured freestone, and the general effect is good, although sombre. Silurian slates are quarried in Scotland in Inverness-shire, Perthshire, and Aberdeenshire; also at Killaloe and some other localities in Ireland.

Devonian slates of a grey colour are worked in Cornwall, at the Delabole and Tintagel quarries, and in Devonshire, in the neighbourhood of Tavistock, at Wiveliscombe and Treborough in Somersetshire, and in other parts of the United Kingdom.

The *carboniferous* flags are quarried for roofing and paving purposes at several places in Yorkshire, Lancashire, and other counties, where carboniferous rocks occur, and are mainly procured from the coal measures. They are of dark-grey colour or black, and are principally used in the neighbourhoods where they are quarried.

There are no true clay slates of later age in Great Britain, but in other parts of the world slates of even tertiary age occur.

Of the clays, used in this country for economic purposes, may be mentioned the china clays or kaolins of Cornwall, which have been formed from the decomposition of the felspathic constituents of granite; the Watcombe clay, which occurs in the trias, and is now used in the manufacture of terra-cotta pottery; the calcareous liassic clays, used for brick-making and burning for lime and hydraulic cement; the various clays of oolitic and neocomian age, some of which are used for brick-making, &c.; the gault, the clays of the Woolwich and Reading beds, and the London clay, all of which are used for bricks; the celebrated Poole clay, dug at Wareham, which belongs to the Bagshot series, and is extensively used for pottery. The clays of the Bovey beds, large quantities of which are annually shipped at Teignmouth, afford good pottery-clays and pipe-clays. There are also many brick-earths and clays of post-tertiary age which are extensively used for brick-making and other purposes. Fuller's Earth is a yellowish, greenish or bluish clayey rock containing about 50 per cent. of silica, 20 per cent. of alumina, 25 per cent. of water, and a little oxide of iron. It chiefly occurs between the Inferior- and Great-Oolite, and in the Lower Greensand. The river-mud in the Medway and at the mouth of the Thames is largely used in the manufacture of Portland cement, after being artificially mixed with chalk and burnt. By the careful levigation of some clays, Dr. John Percy has eliminated minute, but beautifully-developed, crystals of kaolinite.

CALCAREOUS GROUP.
(*Limestones.*)

These, in some cases, consist almost exclusively of carbonates of lime and magnesia (magnesian-limestones or dolomites), while occasionally they are very impure, containing a considerable admixture of sand, clay, and, in some instances, bituminous matter.[1] The British *Silurian* limestones are of comparatively little value, except for lime, locally employed for agricultural purposes. The *Devonian* limestones are, however, extensively used for building and paving, and some of them are well adapted for ornamental purposes on account of the richly coloured mottling and veinings which they frequently exhibit. The carboniferous limestone is largely used for building, and is a very durable stone. Some of the highly fossiliferous beds constitute handsome marbles, the fossils being white, and the rock itself dark grey, or almost black. Good encrinital marble is quarried at Dent in Yorkshire, and the stone is much used for chimney-pieces. The carboniferous limestone often contains bands and nodules of chert. The magnesian limestone of *Permian* age[2] is a very well known, and, when judiciously selected and properly laid, a very durable build-

---

[1] The microscopic structure of the principal English limestones has been described by Mr. H. C. Sorby in his presidential address to the Geological Society, 1879.

[2] Beds of magnesian limestone also occur in the carboniferous limestone series in Derbyshire and elsewhere.

ing-stone. This is well shown in the keep of Conisborough Castle and York Minster, which have both been built of Magnesian Limestone. The Museum of Practical Geology is fronted with this stone, and has stood well. The Houses of Parliament are also built of Magnesian Limestone. It may here be observed that many limestones contain only a very small percentage of carbonate of magnesia, and since magnesia and lime are isomorphous, the amount of magnesian carbonate in limestones may fluctuate from mere traces to a ratio of $CaCO_3$ to $MgCO_3 = 1 : 3$. No sharp line of demarcation can therefore be drawn between the dolomitic limestones and the true dolomites, in which the ratio of $CaCO_3$ to $MgCO_3 = 1 : 1$ giving the percentage composition as $CaCO_3 = 54\cdot35$, $MgCO_3 = 45\cdot65$ for normal dolomite.

The *liassic* limestones ('cement-stones') are argillaceous, and are burnt for hydraulic lime.

The *oolitic* limestones are so numerous and constitute such valuable building-stones that it is only possible to mention a few of those principally employed. These are the Doulting stone, belonging to the inferior oolite, the Bath-stones belonging to the great oolite, of which the chief kinds used are the Box Hill and Corsham Down stones. The Ketton stone, belonging to the Lincolnshire oolites, is an exceedingly valuable building-stone, possessing great tenacity, working freely, and resisting atmospheric influences, even when placed in unfavourable situations. The Ancaster stone, which also occurs in the Lincolnshire oolites, is a less expensive stone, but is very durable. The Portland oolites afford remarkably good stone, which is used very largely, and constitutes one of the most important building-stones in this country. The Purbeck limestones, which, unlike the preceding, are of fresh-water origin, have been used for paving; while, in the upper part of the series, a compact limestone, crowded with fossil shells, of the genus Paludina, is known as Purbeck marble, and has been used for architectural decoration, for some centuries. The Petworth or Sussex marble (Wealden) has also been applied to similar purposes. The oolitic limestones, as a rule, differ structurally from the limestones of older and of more recent date, inasmuch as that they are usually aggregates of little spherical deposits of carbonate of lime, which have formed in concentric crusts around nuclei. These nuclei consist sometimes of a granule of sand, sometimes of the remains of a minute organism. The little spherules are seldom much bigger than a large pin's-head, and they are also cemented together by calcareous matter. The name oolite is derived from the egg-like or fish-roe-like appearance of the stone; but oolitic structure, although characteristic of the limestones of oolitic age, is, however, not exclusively peculiar to them, for well-developed oolitic structure occurs in certain beds of the carboniferous limestone near Bristol, while it is also developed in the coarser pisolites or pea-travertines of recent date. The older limestones, such as those of the Devonian, carboniferous, and Permian formations, are either granular or crystalline-granular, and the latter character is beautifully shown in certain limestones at and near their contact with eruptive rocks. The saccharoid statuary marbles of Italy are good examples of this structure, and, when examined in thin slices under the microscope, are seen to consist of closely aggregated crystalline grains, in each of which polarised light reveals the existence of numerous twin lamellæ, the twinning taking place along planes parallel to the face $-\frac{1}{2}$ R. It seems impossible, in many cases, to say whether this structure in limestones has been due to the metamorphism engendered by the contact or proximity of eruptive rocks, or whether it is owing to other causes, since we find precisely the same structure in the amygdaloids of calcspar which have been infiltered, into the vesicles and crevices in basalts, long after their solidification; we find it in the fossils of the chalk and of other formations, which have not had the opportunity of becoming altered by the presence of intrusive rocks; and we also find it in limestones, at and near their contact with eruptive masses, as already observed. The earthy variety of limestone, chalk, has been stated to consist exclusively of the calcareous tests of foraminifera. This, however, is not always the case. Samples of chalk may sometimes be carefully levigated and examined, and foraminiferal remains may only be detected here and there, the greater part of the matter having merely the character of an ordinary amorphous precipitate; while, again, other samples may be found to consist in great part or almost entirely of the remains of these organisms. Some writers have even gone so far as to express an opinion that nearly all limestones have been formed out of the calcareous remains of foraminifera, corals, &c. In controversion of these statements, Credner, besides giving other good reasons, appeals to microscopic evidence, which shows, he observes,[1] 'that our ordinary compact limestones are by no means always formed of broken and finely-ground organic remains, but rather of little rhombohedra of calcspar.' On the other hand, however, we are bound to admit that the fossils which occur so plentifully in limestones, at all events, represent something more than an insignificant proportion of their bulk, and in some cases seem even to constitute the greater part of the rock.

Cretaceous limestones.—The Kentish rag is mostly a very hard sandy limestone, and contains more or less dark-green glauconite, generally in fine, occasionally in coarse, roundish grains. Glauconite is stated to sometimes form the cementing medium in these rocks, but more or less carbonate of lime is always present in this capacity. By decomposition, the protoxide of iron in the glauconite is converted into peroxide of iron, and the rock, under these circumstances, assumes a reddish-brown tint. According to

[1] *Elemente der Geologie*, Leipzig, 1876, p. 290.

Ehrenberg, the glauconite grains often fill, invest, or replace the tests of foraminifera. These rocks form very durable building-stones. Besides their use in ashlar work they are often laid in irregularly-shaped blocks, giving rise to a honeycomb pattern on the surfaces of the walls built of them. Kentish rag is chiefly quarried at Maidstone, Hythe, and Folkestone, and is extensively used for building in the South of England. It is derived from the Hythe beds.[1] Limestone, either as ordinary chalk or as subordinate beds of compact limestone, represents a considerable part of the cretaceous series of rocks, while most of the cretaceous sandstones are very calcareous. The chalk attains a great thickness in some parts of the kingdom; the lower portion, termed the grey chalk or chalk marl, is generally slightly glauconitic at the base. The upper chalk contains numerous nodules, and occasionally bands of flint, which follow the stratification, although at times vertical bands of flint occur, filling up what once were open fissures. Chalk, besides being largely burnt for lime, is also locally used for building. Certain hard beds occur in the chalk which are better suited for this purpose than the softer material.

Tertiary limestones.—In the British Isles these are but poorly represented. The Binstead limestone, occurring in the Bembridge beds in the Isle of Wight, has, however, been extensively quarried, and has been employed in the construction of some of our early churches. In other parts of the world tertiary limestones often attain great thicknesses, and constitute important building stones. The pyramids, for example, are built of nummulitic limestone.

There are many other interesting tertiary limestones, but want of space precludes any mention of them.

ALTERED SEDIMENTARY SERIES.

With regard to the rocks of this series it is difficult to

[1] An account of the Wealden marbles will be found in Mr. W. Topley's 'Geology of the Weald,' *Mem. Geol. Surv. Eng. & Wales*, p. 368.

say where alteration begins and where it ends, still more difficult in some cases to define the nature of the alteration. It is common for geologists to talk about altered slates where they show the slightest perceptible difference from slates which they regard as normal types, but it is often open to question how far the normal types are really normal, and to what extent the microscopic crystalline constituents of these rocks are to be considered normal or of secondary origin. Von Lasaulx, for example, states a belief that some of the microliths in slates may be referred to hornblende and epidote, while the latest researches of Kalkowsky tend conclusively to show that, in many cases, they are staurolite.[1]

The rocks of the altered sedimentary series may be divided into

    A. Those with no apparent crystallisation.
    B.   „   „ sporadic crystallisation.
    C. Crystalline { *a.* non-foliated.
                         { *b.* foliated and schistose.

Altered sedimentary Rocks with no apparent crystallisation.—In the case of limestones, a crystalline or crystalline-granular condition frequently results from alteration, but sometimes the change simply appears to cause induration, without developing any crystalline structure, as in some of the Antrim chalk, altered by the proximity of basalt. Sandstones, from the character of their constituent particles, can hardly be included under this division of the altered sediments. The alteration which argillaceous rocks undergo without begetting any perceptible crystallisation consists mainly of changes which appear to be of a purely physical character, generally slight and difficult to describe intelligibly. Perhaps the best example of such alteration is to be found in the so-called porcelain jaspers, clays, or shales, which have been baked, either by the combustion of adjacent

---

[1] '*Über die Thonschiefernädelchen.*' *Neu. Jahrb.* 1879.

coal-seams or by the contact or proximity of eruptive rocks. Porcelain jasper has a fused or fritted appearance, a slight gloss, and the different bands or laminæ often assume strongly-marked differences of colour, in which dark green and brick-red sometimes predominate.

Altered sedimentary Rocks with sporadic crystallisation.— In these rocks the development of the crystals is often very imperfect and obscure; in some cases, however, the crystals are distinct and well-developed.

The sporadic crystals which occur in altered limestones are varieties of pyroxene, usually coccolite, hornblende, garnet, sphene, tourmaline, spinel, phlogopite, chlorite, talc, &c., but the limestones themselves, in which these crystals occur, almost invariably have a crystalline structure engendered by metamorphism, and consequently they should rather be placed amongst those altered sedimentary rocks which have a crystalline structure.

The altered slates frequently exhibit sporadic crystallisation. The crystals developed in them are usually silicates of alumina, such as staurolite, andalusite, chiastolite, &c.

Chiastolite slate occurs in the neighbourhood of granitic masses, as in the Skiddaw district in Cumberland, where, according to J. Clifton Ward, 'on approaching the altered area the slate first becomes faintly spotty, the spots being of a somewhat oblong or oval form, and a few crystals of chiastolite appear. Then these crystals become more numerous, so as to entitle the rock to the name of chiastolite slate. This passes into a harder, more thickly-bedded, foliated and massive rock, spotted (or andalusite) schist; and this again into mica schist of a generally grey or brown colour, and occurring immediately around the granite.'[1] The chiastolite slates are mostly of dark grey or bluish-black colour, and contain pale yellowish-white crystals of chiastolite, sometimes more than half an inch in length. These in trans-

---

[1] 'Geology of the Northern Part of the English Lake District.' *Memoirs of the Geol. Surv. England and Wales*, 1876, p. 9.

verse section often show a dark central spot and, occasionally, the chiasmal interpositions which characterise this mineral.

Staurolite slate is a dark micaceous slate containing crystals of staurolite; in some localities passages have been observed from this rock into andalusite slates.

In some altered slates the staurolite is so imperfectly developed that it merely appears in roundish or lenticular knots or lumps, which exhibit no approximation to crystalline form.

The knoten- frucht- garben- and fleckschiefer of German petrologists consist of micaceous slates containing small irregular concretions or little lenticular or ovoid bodies, which, in some cases, may be referred to andalusite, but in many instances they are shown by microscopic examination to be aggregates of small scales of mica, carbonaceous matter, quartz granules, and other constituents of the rocks in which they occur. They are often surrounded by ferruginous stains resulting from decomposition. The determination of the precise character of these bodies is often a matter of considerable difficulty.

Want of space precludes any description of other interesting rocks which belong to this group.

ALTERED SEDIMENTARY ROCKS (CRYSTALLINE).

These rocks may be divided into (*A*) non-foliated, and (*B*) foliated and schistose groups.

A. *Non-foliated Group.*

Under this head come the limestones in which a crystalline structure has been superinduced by the proximity of eruptive rocks. This structure differs, in no essential respect, from that of the crystalline limestones described at page 284. The sporadic crystals, which sometimes occur in these limestones, are coccolite, tremolite, tourmaline, sphene, chondrodite, spinel, garnet, mica, chlorite, &c. Those crystalline limestones which are suitable for ornamental architecture are termed marbles, and many marbles are rocks of this kind, which owe their crystalline character to alteration by intrusive masses; still there are also many in which the crystalline structure is not due to this cause. The term marble is, however, very loosely employed, and may be generally taken to signify any rock which takes a good polish and is employed for decorative or architectural purposes. It is impossible, from want of space, to allude even to the most important marbles.[1]

Quartzite is a compact crystalline-granular aggregate of quartz, either in irregular crystalline grains, or in well-developed crystals. Some quartzites exhibit a schistose structure, which is partly, or wholly, due to the presence of small quantities of mica, the scales of mica lying in the direction of the fissile planes in the rock. These schistose quartzites may therefore be regarded as mica schists poor in mica.

Lydian-stone (basanite, touch-stone, kieselschiefer) is a dark-coloured, generally velvet-black or brownish-black rock. It is an altered sandy slate. Under the microscope, it is seen, in great part, to consist of crystalline grains of quartz mixed with particles of argillaceous, carbonaceous, and ferruginous matter. The percentage of carbonaceous matter is sometimes considerable, and accounts for the extremely black colour of the rock. Lydian-stone is often traversed by small veins of crystalline quartz, and frequently contains a little pyrites.

B. *Foliated and Schistose Group.*

These rocks which are commonly designated crystal-

---

[1] Good accounts of the Italian marbles will be found in the *Official Catalogue of the Exhibition of* 1862, 'Kingdom of Italy,' section v. p. 44; and in *The Mineral Resources of Central Italy*, by W. P. Jervis. London, 1868.

line schists, afford, perhaps, the best-defined instances of metamorphism, in the sense in which that term is usually applied. They are generally characterised by the presence of one of the following minerals, hornblende, mica, chlorite, talc, and occasionally schorl ; but gneiss, described under the granitic rocks at p. 210, has the same constituents as granite, and is in many, if not in nearly all, instances, also an altered sedimentary rock. How far its foliation always represents bedding, is, however, a point which does not yet appear to be fully demonstrated. The term gneiss has been applied to many rocks which have not the same mineral constitution as granite, and which should rather be referred to hornblendic, and other, schists, into which, however, gneiss sometimes passes. There are many varieties of gneiss, of which the most important were alluded to under the granitic rocks at page 210, more on account of their mineral constitution than of their origin, mode of occurrence, and structure, which latter would entitle them to be placed in this group.

Gneiss[1] is a foliated crystalline aggregate of the same minerals which constitute the different varieties of granite, typically, of orthoclase, plagioclase, quartz, and mica. These minerals are arranged in more or less distinct layers or foliæ which are approximately parallel to one another. The mica, especially, forms very distinct, although thin, bands, and it is to this arrangement of the mica that the schistose and often fissile character of the rock is due. Sometimes the mica is a potash, sometimes a magnesian mica, and, at others, both kinds are present. Gneiss varies in colour, the orthoclase in some varieties being red, while in others it is white or greyish. These different rocks are on this account designated red gneiss and grey gneiss, and it has been shown, by analysis of some of the most typical examples, which occur in the neighbourhood of Freiberg, that there is a marked chemical difference between them, the red gneiss containing

[1] See also page 211.

from 75 to 76 per cent. of silica, while the grey variety possesses only from 65 to 66. From these analyses the percentage of the constituent minerals in the two rocks have been deduced as follows :

Red gneiss : orthoclase = 60, quartz = 30 mica = 10 per cent.
Grey gneiss : „ = 45, „ = 25 „ = 30 „

The above represent extremes of variation, while numerous transitional conditions of gneiss exist between them.

In some varieties of gneiss the mica, instead of lying in parallel foliæ, wraps round lenticular aggregates of felspar and quartz, or round crystals of orthoclase. These varieties are termed pseudo-porphyritic gneiss, eye gneiss (*augen gneiss*), wood gneiss, &c.

Oligoclase, dichroite, garnet, micaceous iron, magnetite, and chlorite sometimes occur plentifully in gneiss, so that they impart a distinct character to the rock, which is then accordingly termed dichroite gneiss, magnetite gneiss, &c.

Protogine gneiss.—This rock has already been described at page 212.

Syenitic or *hornblendic gneiss* has the same mineral constitution as syenitic granite. The felspar is, in great part, represented by oligoclase. It is a rock of very extensive occurrence, and passages have been observed from hornblende gneiss into hornblende schist.

Granulite[1] or leptinite is a schistose rock composed of orthoclase or microcline, and quartz, and contains, as a rule, numerous small garnets. When mica is present the rock assumes a more or less schistose structure, and passes over into gneiss-granulite, or gneiss. Schorl is of common occurrence in this rock. The margins of granitic masses sometimes approximate very closely in character to, or are even identical, in mineral constitution, with, granulite. When this is the case the rock must certainly be eruptive, unless it

[1] See also page 210.

can be regarded as an alteration of adjacent sedimentary rocks. At Brazen Tor, on the western margin of Dartmoor, the granite assumes quite a granulitic character, so far as its constituents are concerned, although it shows no foliation or schistose structure, while the eruptive rock which intervenes between it and the sedimentary rocks seems to preclude the idea that its origin is other than eruptive, and that it is anything more than a phase of the granite.[1] The schistose varieties of granulite must, however, be regarded as altered sedimentary rocks.

Erlan rock occurs in mica-slate at Erlhammer, near Schwarzenberg, in Saxony. It is a fine-grained or compact aggregate of garnet, albite, and quartz, and occupies a place intermediate between granulite and garnet rock.

Porphyroid.—Under this name are included certain altered sedimentary schistose rocks, consisting of a fine-grained matrix of felspar and quartz, and containing a large quantity of a sericite-like or micaceous mineral. Within this matrix lie numerous crystals of felspar (orthoclase or albite) and rounded grains or crystals of quartz. The porphyroids occur interbedded with other old sedimentary rocks.

Sericite-Schist.—This is a schistose rock closely allied to the porphyroids, and consists of sericite, fragments of quartz, albite, and usually more or less chlorite and mica. There are two varieties, the green and the red, which differ somewhat in composition, the former containing much albite and little or no mica, while the latter contains mica and very little albite. In some cases the rock is rich, in others poor, in quartz. The very fine-grained, compact varieties in which the constituent minerals cannot be distinguished by the naked eye, are termed sericite-slates or sericite-phyllites. Occasionally augite may be detected in some varieties of this rock, which, in such cases, are evidently allied to the diabase-schists or tuffs.

[1] 'The Eruptive Rocks of Brent Tor,' *Memoirs of Geol. Surv.* 1878.

Mica-Schist is an aggregate of mica and quartz. The relative amounts of these two minerals varies considerably, some varieties of the rock consisting almost wholly of mica, while others are composed almost wholly of quartz, and contain only a very small proportion of mica. The latter approximate and pass into quartz-schist. All of these rocks have a schistose structure due to the parallel arrangement of the crystals and scales of mica. The mica in these schists is sometimes silvery-white potash-mica, sometimes dark magnesian-mica, but the former is by far the most common. The quartz is in small grains, often of a flattened or lenticular form, and it is very usual for the quartz and mica to constitute alternating parallel layers, so that the rock exhibits more or less distinct foliation. The percentage of silica in mica-schists varies according to the amount of quartz which is present, the extremes of fluctuation being between 40 and a little over 80 per cent. Garnets are of common occurrence in these rocks, and numerous other minerals are met with as accessories, such as tourmaline, hornblende, kyanite, staurolite, felspars, epidote, chlorite, talc, magnetite, pyrites, specular-iron, gold, and graphite. When certain of these minerals preponderate, the rocks pass over into schists of a different character, to which special names are given. In some of these rocks calcspar or dolomite partly or wholly replaces the quartz and films of argillaceous matter; chlorite and sericite are also present at times. Such rocks are known as calcareous mica-schist, calcareous chlorite-schist, &c.

Itacolumite is a somewhat schistose, micaceous sandstone consisting of granules of quartz and scales of mica, talc, chlorite, &c. The rock is flexible, when in thin slabs. It occurs in N. America and Brazil, and contains, amongst other minerals, gold and diamonds.

Chlorite-Schist usually occurs interbedded with gneiss and with other metamorphic rocks. It is of a green or greenish-grey colour, and, as a rule, consists in great part of scales of chlorite closely matted together. Quartz, however, is usually present, and felspars, mica, and talc are also of common occurrence. Sometimes it exhibits a fine-grained and evenly-cleavable slaty character (chlorite-slate), but the coarser schistose varieties usually have a somewhat imperfect, or irregular and wavy, fissile structure. The chlorite-schists are frequently rich in accessory minerals, among which may be cited mica, hornblende, actinolite, schorl, epidote, sphene, corundum, garnet, rutile, specular-iron, magnetite, pyrites, copper-pyrites, and gold.

Talc-Schist consists of scales of talc with a small admixture of quartz. It is a greenish, greyish-white, or yellowish-white rock, very soft, and has a smooth greasy feeling when rubbed with the fingers. Chlorite and mica are often present, and sometimes a little felspar. The rock contains as accessories many of the minerals which occur in chlorite-schist, in addition to which may be mentioned asbestus, hydrargillite, fahlunite, gahnite, chlorospinel, &c.

Potstone is a soft, sectile, greenish-grey rock, composed of chlorite, talc, and serpentine, used in Italy for the manufacture of cooking-pots.

Schorl-Schist is mainly composed of schorl and quartz, and often shows well-marked foliation, the greyish crystalline-granular rock being traversed by approximately parallel thin black bands of schorl which are often more or less contorted. Cassiterite very commonly occurs in this rock, also mica, chlorite, felspars, topaz, and arsenical pyrites.

Schorl rock is a granular, non-foliated rock of similar mineral constitution.

Amphibolite-, Hornblende-Schist consists of hornblende and quartz. The latter mineral is often present only in small quantities. It is a dark greenish-grey or iron-grey rock, and when quartz is absent, and it consists exclusively or almost exclusively of hornblende, it is then termed amphibolite. Some of these amphibolites may be regarded as truly eruptive rocks. The hornblende is sometimes granular, at others in radiating prisms or in a fibrous condition. Magnetite is common as an accessory, and felspars, garnet, biotite, epidote, and pyrites are also met with, although they are always subordinate constituents.

Actinolite-Schist may be regarded as a variety of hornblende-schist. The constituents are actinolite and quartz.

For an account of the artificial production of schistose structure, see Daubrée's 'Etudes Synthétiques de Géologie Expérimentale,' p. 407.

COARSE FRAGMENTAL ROCKS.

These may be divided into Breccias and Conglomerates. They consist of materials derived from the waste of various rocks and are made up of fragments either angular, or sub-angular, or of rounded, waterworn pebbles or boulders. Similar but much finer material constitutes many of the sedimentary rocks, and it is merely in the size of the fragments that many sandstones, grits, &c., differ from breccias and conglomerates, the former being as much entitled to be placed among the fragmentary or clastic rocks as the latter. This view is very clearly represented by Naumann's classification of the clastic rocks,[1] which he divides into the psephitic (from ψῆφος, a small stone); the psammitic (from ψάμμος, sand); and the pelitic (from πηλός, mud). The psammites and pelites of the two last groups are respectively represented by the various sandstones, arkose, &c., and by the tuffs which have already been described in conjunction with the rocks from which they have been derived; so that it only remains to describe the breccias and conglomerates which constitute the psephitic division of the clastic rocks, although the coarse materials of the latter are often mixed with or cemented by psammitic and pelitic matter.

BRECCIAS.

The breccias differ from the conglomerates in consisting

[1] So named from κλαστός, broken.

of angular or sub-angular fragments, instead of rounded water-worn pebbles or boulders; but it often happens that these coarse clastic rocks have a mixed character; consisting partly of angular and comparatively unwater-worn, and partly of rounded water-worn materials. To such rocks the name breccio-conglomerate is sometimes given. The fragments of which breccias are composed are usually large enough to permit the recognition of their lithological character. They are often derived from various sources, fragments of sandstone, quartz, jasper, and various eruptive rocks being common, while occasionally they consist of limestone fragments, or of broken pieces of bone, as in some of the bone-breccias which occur in cave-deposits. The generally angular character of breccias indicates that they have been formed at or near the spots where they occur and that their materials have never travelled far. They may be formed from the superficial disintegration of rocks in their immediate vicinity; they may represent talus and rubbish heaps which have been subsequently cemented by the infiltration of calcareous, siliceous, or ferruginous matter; they may result from the breaking in or 'creep' of rocks with which soluble deposits have been interstratified and subsequently dissolved out, as in the case of the *haselgebirge*, occurring in the Northern Alps, where, from the removal of underlying salt beds, a breccia has been formed, consisting of fragments of various rocks imbedded in a matrix of clay.

Breccias may also result from the breaking-away of small fragments from the sides of fissures (friction breccias) through which dykes of eruptive rock have been subsequently injected, or into which mineral matter in solution may have subsequently filtered. The latter condition is a very common one, and may be well observed in some metalliferous lodes. In such cases it is not unusual to find the fragments enveloped in successive deposits of different mineral character. The accumulations of angular and sub-angular ice-worn and scratched stones which constitute the moraines of glaciers, and the angular unworn rubbish dropped by the melting of icebergs, may also give rise to deposits which may be regarded as unconsolidated breccias. Sometimes, also, they are formed at the bottom of lava-flows, the once molten rock having caught up and enveloped fragments of various kinds and sizes; and, finally, coarse volcanic ejectamenta may also constitute rocks which may justly be described as volcanic breccias or agglomerates.

Some breccias, when polished, are well adapted for ornamental purposes. They occur in formations of different ages, and have an especial geological interest as indicating either the local disintegration of older rocks, or the transport of materials from distant countries by glacial agency.

CONGLOMERATES.

These are composed of pebbles or boulders which have been either carried by rivers, or washed about on shores, and consequently rounded by attrition. The process of rounding is constantly going on in the beds of our rivers and along our sea-coasts; and the beaches which are formed along our shores would, if cemented, become true conglomerates. The materials composing conglomerates are, like those of breccias, of very variable character, being derived from many different sources, but they are mostly formed from rocks of considerable hardness, since softer fragments become totally disintegrated by constant trituration and abrasion. The beaches now forming on the south-eastern coasts of England are in great part composed of pebbles which have resulted from the wear and tear of flints derived from the chalk-cliffs. The puddingstone of Hertfordshire consists of flint pebbles, held together by a siliceous cement. The nagelfluhe, formed on the northern flanks of the Alps, consists, in great part, of limestone fragments, partly mixed with fragments of quartz, granite, gneiss, &c. The new-red conglomerate (Keuper) consists mainly of pebbles and boulders of carboniferous limestone, usually cemented by dolomitic matter, whence it is also called dolomitic conglomerate. The pebble beds of the Bunter are formed to a large extent of pebbles of quartz and quartzite, in a matrix of sandstone. Conglomerates occur at the base of the old-red-sandstone in Caithness and elsewhere. These contain not merely pebbles, but large boulders, and the deposit is considered by Prof. Ramsay to be of glacial origin. Conglomerates are sometimes used as building-stones. The dolomitic conglomerate of Triassic age which occurs in the Mendips consists of pebbles and small boulders derived from old-red-sandstone and carboniferous rocks and the cementing material is often, but not invariably, dolomitic. Conglomerates have a special geological interest, inasmuch as they usually represent old sea-beaches, and consequently indicate the former existence of coast lines. There, are, however, instances in which they may not represent littoral deposits.

CALCAREOUS TUFAS.

Calcareous tufa, travertine, pisolite, osteocolla, &c., are deposits formed by the chemical precipitation of carbonate of lime from waters holding bicarbonate of lime in solution. Deposits of this kind are generally formed in the valleys of limestone districts. Any foreign bodies which occur in the solution from which the precipitation takes place, become externally incrusted, just as kettles and boilers become furred internally with carbonate of lime. Successive deposits are thus formed, and the result is a light and often spongy rock, in which more or less distinct layers represent the successive deposits. Twigs, leaves, and other objects become, in this manner, incrusted with carbonate of lime; and a small trade is carried on at Matlock, Knaresborough, and elsewhere, by submitting natural and artificial objects to the incrusting influence of the waters of these petrifying-wells, as they are termed. The variety of tufa named osteocolla consists of calcareous deposits around twigs and mosses, while pisolite is composed of little pea-like spherical concretions of carbonate of lime around small nuclei. The oolitic limestones have been formed to some extent in a similar manner, although under very different circumstances; the latter representing marine deposits, and consisting largely of comminuted organic remains, while the ordinary calcareous tufas are usually formed in valleys, and constitute, as a rule, deposits of very limited extent. In Italy, however, some extensive deposits of travertine occur, especially at Tivoli, where the waters of the Anio have formed beds of tufa four or five hundred feet in thickness.[1]

Calcareous tufa is sometimes used as a building-stone, and appears to be very durable, even in old Roman edifices.

Siliceous Sinters.

These are rocks formed by the deposition of silica from the waters of hot springs and geysers. Geyserite is a snow-white siliceous sinter of this kind which occurs incrusting the pipes of geysers and forming tolerably thick deposits on the adjacent ground over which the water of the geysers is ejected. The deposits of siliceous sinter at Rotomahana, near Lake Taupo, in New Zealand, are perhaps the most wonderful in the world. At this place the thermal waters charged with silica in solution flow down the hill-sides, forming snow-white terraces of siliceous sinter. The influence which thermal waters, holding silica in solution, have exerted upon many of our older rocks, is a question which well deserves the attention of petrologists.

MINERAL DEPOSITS CONSTITUTING ROCK-MASSES.

Rock-salt, in some districts, constitutes deposits of great thickness: coal also forms beds or seams of variable thickness in the carboniferous series, and in rocks of oolitic and miocene age in certain countries.

[1] Lyell's *Principles of Geology*, 9th edition, p. 244.

Gypsum sometimes forms beds of considerable thickness. Iron ores occasionally occur in large masses, as in the case of the Pilot Knob, Missouri, which consists almost wholly of hematite, of some of the large deposits of this ore at Norberg and Langbanshytta in Sweden, Gellivara in Lapland, and Santander in Spain, and of the extensive deposits of limonite and clay-ironstone which occur in various parts of the world. Felspars, such as labradorite, sometimes constitute, by themselves, rock-masses of considerable thickness and extent. In Labrador, rocks consisting chiefly of labradorite form large and important beds. Various metallic ores at times constitute lodes and beds of considerable magnitude.

The deposits of cinnabar in Spain, of zinc ores in New Jersey, U.S., and in the Harz; of copper in the Lake Superior district, and of other minerals of the heavy metals which occur more or less plentifully in all parts of the world, the deposits of rock-salt in Russia, Poland, Gallicia, Germany, Austria, Spain, England, Algeria, Abyssinia, and in various parts of America, the phosphatic beds met with in certain formations, and the important occurrences of coal, lignite, bituminous schists, asphaltum, and petroleum, are matters which can be studied in most geological, mineralogical, and mining books.

Ice, which must be regarded as a rock, occurs in thick and extensive sheets over the Arctic and Antarctic regions, while perennial snow exists at great elevations, even at the equator. Ice is sometimes interstratified with sands, &c., as in some parts of Siberia. Deception Island, in New South Shetland, lat. 62° 55′ S., is principally composed of alternate layers of volcanic ashes and ice, and similar alternations of beds have been observed on Cotopaxi, while a large glacier has been discovered by Gemmellaro, beneath a lava-stream, at the foot of the highest cone of Etna. The latter phenomenon is described in Lyell's 'Principles of Geology,' and the author attributes the preservation of the glacier to a mass of snow having been covered by a badly conducting layer of volcanic ash, scoriæ, &c., prior to the eruption of the lava which now caps the glacier. Ice exhibits stratification in the upper portions of glaciers, but the motion of the latter in their descent to lower levels gradually obliterates these traces of bedding. Glaciers and ice-sheets, in creeping over the subjacent rocks, scrape them into smooth, hummocky forms, termed *roches moutonnées*, by means of the stones, gravel, &c. which get imprisoned between the ice and the rock-surface over which it moves. The phenomena of glaciation will be found fully discussed in many books and papers which have been published on the subject during the last ten or fifteen years.

APPENDIX.

A.

SINCE this book was written, Mr. Watson, of Pall Mall, has, at my suggestion, constructed a microscope specially suited for petrological work.

This instrument is in one respect decidedly preferable to the microscopes commonly made on the Continent, inasmuch as it is supported upon trunnions, like the ordinary English microscopes, and consequently allows of any inclination of the working part of the instrument. The foot, up to the trunnion-bearings, is cast in one piece, after the model of Ross, Crouch, and other well-known makers. Upon this a brass limb is supported. The limb, below the trunnions, is cylindrical, and carries an ordinary mirror with a jointed arm. The limb, above the axis, describes such a curve as is most convenient for lifting and carrying the instrument, without incurring any risk of strain to the working parts. The upper portion of the limb is planed to receive the rackwork, which constitutes the coarse adjustment, in a manner similar to that employed in the construction of the Jackson-pattern microscopes, as in the instruments by Beck. The tube or body carries the rack, and, by it, is moved against these planed surfaces, for focussing. At the lower end of the tube, immediately above the thread which carries the objective, a slot is cut to receive a Klein's quartz plate. The quartz plate is in a small brass mount, which fits this slot, and can be removed from the instrument at pleasure. At such times a revolving collar can be turned over the aperture. The eye-piece, at the upper end of the tube, is made with a disc, about $1\frac{1}{2}$ inch diameter, having its outer edge divided, and, immediately above this, a similar disc, connected with the eye-piece analyser, revolves with an index, so that the analyser can be set in any desired position, or the amount of rotation imparted to it can be recorded. The eye-piece is also furnished with crossed cobwebs for centering. A space is left between the bottom of the analyser and the eye-glass, sufficient to permit the introduction of a plate of calcspar for stauroscopic examinations, and an eye-piece-fitting, without lenses, is also supplied, so that the instrument can, by the superposition of convergent lenses over the polariser, be used for viewing the systems of rings and interference crosses, presented by crystals when examined by convergent, polarised light. The polarising prism is mounted upon a movable arm, beneath the stage, and carries a graduated disc, so that it can be set in any desired direction, or be instantly displaced when ordinary illumination is requisite. The stage of the microscope is circular, and can be rotated and centered. It is divided on silver to half degrees, and a vernier is attached to the front of the stage for goniometric purposes. It has also two rectangular, divided lines, to serve as a finder. The centring is effected by two screws working against a spring on the opposite side. These screws enable the observer to centre the instrument for any objective. They are conveniently situated, so as not to be liable to derangement during the ordinary manipulation of the instrument. The fine adjustment carries a divided head, for the approximate measurement of the thickness of sections. The body of the microscope can be clamped in any position by a lever, attached to the right trunnion.

In the instrument which I have examined the adjustments work very smoothly. The thread for the objectives is of the, now, almost universal gauge, so that any English objectives may be used. Foreign ones can also be employed, by means of an adapter.

The engraving on page 306, for which I am indebted to Mr. Watson, shows the general plan upon which the instrument is constructed. The smaller figures represent the quartz plate, the calcspar plate, and a section of the polariser and its fittings.

The calcspar plate supplied with this instrument is used for the same purpose as that which is fitted in the stauroscope described at p. 81. The quartz plate affords a delicate means of determining the maxima of extinction in doubly refracting minerals, since, so long as an axis of elasticity fails to coincide with the principal section of one of the crossed Nicols, the uniform tint of the field is disturbed, and is only restored when coincidence is effected. It also affords a valuable means of detecting particles of isotropic matter in rock-sections.

B.

Tables showing the angles made by the directions of extinction in triclinic felspars with the edge E E formed by the faces oP and $\infty \breve{P} \infty$.

(According to Max Schuster.[1])

On cleavage plates taken parallel to oP.

Albite	+ 4° to + 3°
Intermediate series	+ 2° ,, + 1°
Oligoclase	+ 2° ,, + 1°
Andesine	− 1° ,, − 2°
Labradorite	− 4° ,, − 5°
Bytownite	−16° ,, −18°
Anorthite	−38°

On cleavage plates taken parallel to $\infty \breve{P} \infty$.

Albite	+18°
Intermediate series	+12°
Oligoclase	+ 3° to + 2°
Andesine	− 4° ,, − 6°
Labradorite	−17°
Bytownite	−29°
Anorthite	−40°

The directions of extinction termed positive and negative are indicated in the following figure.

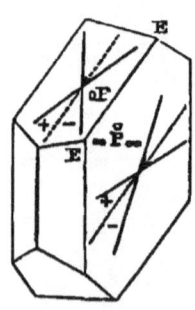

Orthoclase and sanidine may be distinguished from microcline, in plates cleaved parallel to the basal plane, by the maximum extinction in the former occurring when the edge formed by the faces oP and $\infty P \infty$ coincides with the principal section of one of the crossed Nicols, while in microcline maximum extinction takes place when the edge formed by oP and $\infty \breve{P} \infty$ is inclined about 15° to the principal section of one of the crossed Nicols.

[1] From Vol. lxxx. Sitzb. der k. Akad. der Wissensch. Pt. 1. July 1879.

C.

Books consulted in the preparation of this Work.

Ansted, D. T. 'Elementary Course of Geology and Mineralogy.' London, 1850.
Ansted, D. T. 'Applications of Geology to the Arts and Manufactures.' London, 1865.
Blum, J. R. 'Handbuch der Lithologie oder Gesteinslehre.' Erlangen, 1860.
Bořicky, E. 'Petrographische Studien an den Basaltgesteinen Böhmens.' Prag, 1874.
Bořicky, E. 'Petrographische Studien an den Phonolithgesteinen Böhmens.' Prag, 1873.
Bořicky, E. 'Petrographische Studien an den Melaphyrgesteinen Bohmens.' Prag, 1876.
Bořicky, E. 'Elemente einer neuen chemisch-mikroskopischen mineral- und Gesteins-Analyse.' Prag, 1877.
Brewster, D. 'Optics' ('Cabinet Cyclopædia'). London, 1831.
Bristow, H. W. 'Glossary of Mineralogy.' London, 1861.
Bryce, J. 'Geology of Arran and other Clyde Islands.' Glasgow and London, 1872.
Coquand, H. 'Traité des Roches.' Paris, 1857.
Cotta, B. von. 'Rocks Classified and Described.' Translation by P. H. Lawrence. London, 1866.
Credner, H. 'Elemente der Geologie.' Leipzig, 1876.
Dana, E. S. 'Text-Book of Mineralogy.' 2nd edition. New York, 1878.
Dana, J. D. 'System of Mineralogy.' Fifth Edition. London and New York, 1871.
Daubrée, A. 'Etudes Synthétiques de Géologie Expérimentale.' Paris, 1879.
De la Beche, H. 'How to Observe—Geology.' London, 1836.
De la Beche, H. 'Report on the Geology of Cornwall, Devon, and West Somerset.' London, 1839.
De la Beche, H. 'Researches in Theoretical Geology.' London, 1834.
De la Fosse. 'Nouveau Cours de Minéralogie.' Paris, 1858-62.

Delesse, A. 'Recherches sur l'Origine des Roches.' Paris, 1865

Delesse, A. 'Recherches sur les Pseudomorphoses.' 'Annales des Mines,' XVI. Paris, 1859.

Delesse, A. and De Lapparent. 'Révue de Géologie.' Paris.

Descloizeaux, A. 'Manuel de Minéralogie.' Paris, 1862.

Fischer, H. 'Kritische mikroskopisch-mineralogische Studien.' Freiburg, 1869 and 1871.

Ganot. 'Elementary Treatise on Physics.' Translation by Atkinson. London, 1863.

Green, A. H. 'Geology for Students.' Part I. London, 1876.

Hunt, R. 'Mineral Statistics of the United Kingdom, 1858. Part II. Embracing Clays, Bricks, &c., Building and other Stones.' (Mem. Geol. Surv.) London, 1860.

Jannetaz, E. 'Les Roches.' Paris, 1874.

Judd, J. W. 'Geology of Rutland, &c.' (Mem. Geol. Surv.) London, 1875.

Jukes, J. B. 'Student's Manual of Geology.' Edinburgh, 1862.

Kengott, A. 'Elemente der Petrographie.' Leipzig, 1868.

Kinahan, G. H. 'Handy-Book of Rock Names.' London, 1873.

La Vallée Poussin and Rénard, A. 'Mémoire sur les Roches dites Plutoniennes de la Belgique et de l'Ardenne Française.' (Acad. Royale des Sciences de Belgique.) Bruxelles, 1876.

Lasaulx, A. von. 'Elemente der Petrographie.' Bonn, 1875.

Lommel, E. 'The Nature of Light, with a General Account of Physical Optics.' London, 1875.

Lyell, C. 'Principles of Geology.' 9th Edition. London, 1853.

Lyell, C. 'Student's Elements of Geology.' London, 1871.

Macculloch, J. 'A Geological Classification of Rocks.' London, 1821.

Möhl, H. 'Die Basalte und Phonolithe Sachsens' (Nova Acta). Dresden, 1873.

Pereira, J. P. 'Lectures on Polarised Light.' London, 1843.

Phillips, J. 'Manual of Geology.' London and Glasgow, 1855.

Pinkerton, J. 'Petralogy.' London, 1811.

Playfair, J. 'Illustrations of the Huttonian Theory of the Earth.' Edinburgh, 1802.

Ramsay, A. C. 'Physical Geology and Geography of Great Britain.' 3rd Edition. London, 1872.

Rénard, A. 'Mémoire sur la Structure et la Composition Minéralogique du Coticule.' (Acad. Royale des Sciences de Belgique.) Bruxelles, 1877.

Rosenbusch, H. 'Die Steiger Schiefer und ihre Contactzone an den Granititen von Barr-Andlau und Hohwald.' Strassburg, 1877.

Rosenbusch, H. 'Mikroskopische Physiographie der petrographisch wichtigen Mineralien.' Stuttgart, 1873.

Rosenbusch, H. 'Mikroskopische Physiographie der massigen Gesteine.' 1877.

Roth, J. 'Die Gesteins-Analysen.' Berlin, 1861.

Scrope, G. P. 'Volcanos of Central France.' London, 1858.

Sorby, H. C. Presidential Address to Geol. Soc. 1879.

Smith, C. H. 'Lithology, or Observations on Stone used for Building.' London, 1845.

Spottiswoode, W. 'Polarisation of Light' (Nature Series). London, 1874.

Tyndall, J. 'Six Lectures on Light.' Delivered in America. London, 1875.

Tyndall, J. 'Notes on a Course of Nine Lectures on Light.' Delivered at the Royal Institution. London, 1872.

Vogelsang, H. 'Die Krystalliten.' Bonn, 1875.

Ward, J. C. 'Geology of the Northern Part of the English Lake District.' (Mem. Geol. Surv.) London, 1876.

Whitaker, W. 'Geological Record.' London.

Woodward, C. 'Familiar Introduction to the Study of Polarised Light.' London, 1861.

Woodward, H. B. 'Geology of England and Wales.' London, 1876.

Woodward, H. B. 'Geology of East Somerset and the Bristol Coal-Fields.' (Mem. Geol. Surv.) London, 1876.

Zirkel, F. 'Lehrbuch der Petrographie.' Bonn, 1866.

Zirkel, F. 'Microscopic Petrography.' (U. S. Geol. Exploration of Fortieth Parallel.) Washington, 1876.

Zirkel, F. 'Mikroskopische Beschaffenheit der Mineralien und Gesteine.' Leipzig, 1873.

Zirkel, F. 'Untersuchungen über mikroskopisch Zusammensetzung und Structur der Basaltgesteine.' Bonn, 1870.

Also numerous papers published in the 'Quarterly Journal of the Geological Society,' the 'Geological Magazine,' 'Leonhard's Jahrbuch für Mineralogie,' 'Zeitschrift der deutschen geologischen Gesellschaft,' 'Annales des Mines,' and numerous other publications.

Among these, books and papers by the following authors have been specially consulted:

Allport, S.	Judd, J. W.
Bonney, T. G.	Kengott, A.
Bořicky, E.	Lasaulx, A. v.
Cotta, B. v.	Lyell, Sir C.
Dana, E. S.	Phillips, J. A.
Dana, J. D.	Rénard, A.
De la Beche, Sir H.	Rosenbusch, H.
Delesse, A.	Sorby, H. C.
Forbes, D.	Zirkel, F.

INDEX.

ACID, sub class of eruptive rocks, 34, 177
Actinolite, 131
— schist, 295
Aden, globular silica in quartz-trachyte of, 152
Age of strata, determined by means of fossils, 30
Aislaby sandstone, 280
Albite, 97
— type of twinning, 98
Alderley Edge, sandstone of, 279
Allport, S., 107, 122, 152, 182, 190, 193, 197, 245, 260
Altered conditions of pyroxene, 126
— eruptive rocks, 269
— sedimentary series, 289
Amazon stone, 97
Amianthus, 131
Amorphous matter, 170
Amphibole, 127
Amphibolite, 298
Amphigene, 108
Analcime, 159
Anamesite, 252
Andalusite, 143
Andesine, 99
Andesite, 234
Anisotropic substances, 77
Anorthite, 98
Ansted, 277
Anticlinal, term defined, 21
Antrim, chalk of, 290
— keuper sandstone of, 280
Apatite, 145
Aplite, 211
Aqueous rocks defined, 15
Aragonite, 149
Arbroath sandstone, 277

Arenaceous limestone, 20
— sedimentary rocks, 276
Argillaceous limestone, 20
— rocks, 282
Arkose, 280
Arran, pitchstone of, 196
Arranging rock specimens, 39
Asbestus, 131
Ascension, obsidian of, 181
Ashes, volcanic, 267
Asparagus stone, 147
Aubuisson, 209
Augen-gneiss, 295
Augite-andesite, 236
— porphyrite, 240
— syenite, 218
Aussig, tridymite from, 152
Auvergne, domite of, 226
Axes of elasticity, 78
— optical, 78
Axiolitic structure in vitreous rocks, 184

BAGSHOT sands, 281
Banded structure in vitreous rocks, 181
Bangor quarries, 283
Bannisdale slates, 283
Bargate stone, 280
Basalt, 252
— classification of, 253
— columnar structure of, 258
— diagram of deviations from, 261
Basaltite, 252
Basanite, 293
Basic rocks, 34
— sub-class of rocks, 177
Basis, term defined, 168
Bath bricks, 281

Bath stone, 286
Baveno type of twinning, 92
Beale, Lionel S., 59
Beale's reflector, 49
Belonites in pitchstone, 190
Berkum, near Bonn, sanidine-rhyolite of, 224
Binstead limestone, 289
Biotite, 135
Blowpipe apparatus, 45
Bobenhausen, tachylyte of, 201
Bombs, volcanic, 267
Bonney, T. G., 182, 193, 211, 260, 262, 270
Bořický, E., method of analysis, 100
— — 120, 125, 132, 229, 253, 260
Bouteillenstein, 187
Box Hill stone, 286
Bradford flags, 278
Bramley Fall stone, 278
Brazen Tor, granulitic rock of, 296
Breccias, 299
— defined, 17
Breislackite, 125
Breithaupt, 97
Brent Tor, schalstein near, 248
— — volcanic ejectamenta of, 268
Brewster, 142, 150, 159
Brezina, stauroscope of, 83
Bristol Coal-Field, sandstones of, 278
Buch, L. von, 171, 234
Bunsen, 250
Bunter sandstones, 279
Buschbad, perlite of, 183
Byssolite, 131

Index

CABINETS, 43
— Caithness flagstone, 277
Calabria, mica syenite of, 218
Calc-aphanite, 247
Calcareous rocks, 285
— sandstone, 19
— tufa, 18, 302
Calcite, 148
Calcspar, 148
— fluid inclosures in, 149
— twinning of, 149
Cambrian sandstones, 276
— slates, 283
Canada, gneissic syenite of, 218
— moroxite of, 147
Camera lucida, 49
Canary islands, eutaxite of, 232
Cancrinite, 108
Cape Wrath, amphibolite schist of, 218
Carbonic acid, action of, upon certain rocks, 29
Carboniferous limestone, 285
— sandstone, 278
— slates and flags, 284
Carlow flags, 278
Carlsbad type of twinning, 92
Cassiterite, 148
Craigleith sandstone, 278
Cast-iron plate for grinding sections, 61
Centering 57
Chabasite, 160
Chalcopyrite, 157
Chalk, 289
Chalk of Antrim, 290
Chert, 285
Chiastolite, 143
— slate, 291
China clay, 272, 284
— — formation of, 30
Chips for section-cutting, 60
Chlorite, 136
— in quartz, 151
— schist, 297
Chrysolite, 116
— pseudo, 187
Classification of rocks, 174
Clastic rocks, 299
Clays, 17, 282
— of Bovey Beds, 284
— &c., washing of, 73
Cleavage, cause of, 35
— in rocks, 12
Cleavages, diagrams of, 173
— of minerals, 171

Cliffs and escarpments, 27
Clinkstone, 228
Clinochlore, 137
Clip-lens, 44
Coal, preparation of section of, 71
Coast lines affected by relative hardness of rocks, 27
— — changes in, 11
Collecting rock specimens, 39
Columnar structure of basalt, 258
— — in phonolite, 233
Collyweston slates, 283
Condensers, 49
— achromatic, 50
Conglomerate, 17, 19, 297
— dolomitic, 302
— phonolite, 233
Coniston flags, 283
Contortion of strata, 21
Cookworthy, Wm., 273
Copper pyrites, 157
Corals, 18
Cordier, 209
Cordierite granite, 210
Cornubianite, 213
Corriegills, Arran, pitchstone of, 197
Corsham Down stone, 286
Coticule, 142
Cotopaxi, 304
Cotta, B. von, 219, 220
Credner, H., 206, 288
Cretaceous sandstones, 280
Cross-hatching in orthoclase, 94
Crust of the earth, dislocations of, 11
Crypto-crystalline matter, 170
Crystalline eruptive rocks, 202
— limestones, 292
Crystallites, 160
— in obsidian, 187
Cubic system, cleavages in, 171
Czertochin, Bohemia, tachylyte of, 202

DACITE, quartzose, 234
Damascened structure in vitreous rocks, 181
Damourite, 134
Dana, J. D., 90, 117, 121, 131, 147, 257
Darwin, C., 36, 37, 212

Dathe, J. F. E., 87, 245
Daubeny, 37
Daubrée, A., 299
Definition of the term rock, 6
De la Beche, Sir H., 37
De la Fosse, 78   [269
Delesse, A., 130, 171, 221,
Dent marble, 285
Denudation, 25
Descloizeaux, 96, 97, 120, 122, 135, 150, 159, 160
Devitrification, 185
— of pitchstones, 198
Devonian limestones, 285
— sandstones, 277
— slates, 284
Devonshire, schalstein of, 248
Diabase, 244
— amygdaloidal, 247
— aphanite, 247
— porphyrite, 239
— schist, 247
Diabasmandelstein, 247
Diallage, 124
— andesite, 234
Diallagoid augite, 132
Dimetian basalts, 262
Dinorwig quarries, 283
Diorite, 241
— porphyrite, 238
Dip, term defined, 21
Disthene rock, 263
Disturbances of the earth's crust, 9
Dolerite, 252
Dolomite, 149, 285
Dolomitic conglomerate, 302
Domite, 226
Doulting stone, 286
Drachenfels, tridymite from, 152
Dressel, 114
Dundee sandstone, 277
Dunite, 265
Durocher's theory, 34

EARTHQUAKES, 11
Eifel, leucitophyrs of, 256
Eisenglimmer, 155
Eklogite, 263
Elba, specular iron of, 155
Elæolite, 108
Elevation and subsidence of land, 11
Elvan, 204, 210
Encrinital marble, 285
Enstatite, 120
Epidote, 127, 139
Erlan, 296

Index. 317

ERU

Eruptive rocks, general characters of, 32
Eruptive rocks, vitreous, 177
Escarpments and cliffs, 27
Estimation of amount of waste by denudation, 24
Etna, ashes of, 267
— glacier on, 304
— lavas of, 255, 267
Eulysite, 263
Eurite, 209, 214
Eutaxite, 233
Eye-pieces, 52

FALSE bedding, 15
Faroe Isles, heulandite of, 160
Faults, origin of, 13
Felsi-dolerites, 34
Felsitic matter, 167
— pitchstone, 197
Felsite, 168, 213
Felspars, 86
— decomposition of, 30
Felspar-porphyries, 209
— rocks, 304
Felstone, 209, 214
Ferrite, 167
Ferruginous sandstone, 19
Filiform condition of vitreous lavas, 186
Finders, 51
Fingal's Cave, 258
Fire-stone, 281
Flags, 282
— Silurian, 283
Fleckschiefer, 292
Flexure of strata, 21
Flints, 281
Fluid inclosures, 164
Fluxion structure, 163
Foliated rocks, 293
Foliation, 36
Foraminifera, 18
Forbes, David, 2
Forces operating in the interior of the earth, 10
Formation, term defined, 30
Fossiliferous rocks defined, 15
Fossils indicative of conditions under which rocks have been deposited, 15
Fossils replaced by pyrites, 157
Foster, C. Le Neve, 212
Fragmental rocks, 299
Freiberg, gneiss of, 294

HAP

Friction breccias, 300
Fritsch and Reiss, 233
Fritzgärtner, M. G. R., 146
Fruchtschiefer, 292
Fuchsite, 134
Fuller's earth, 284
Fusion of vitreous matter, 179

GABBRO, 249
Garbenschiefer, 292
Garnet, 140
— rock, 262
Gemmellaro, 304
Geyserite, 303
Geysers, 33, 303
Giant's Causeway, basalt of, 258
Glass inclosures, 165
— natural, 170
Globular silica, 152
Globulites, 161
Gneiss, 212, 294
— augen, 295
— foliation in, 212
— protogine, 212, 295
— syenitic, 295
Gneissic syenite, 218
Goniometers for microscopes, 53
Granite, 202
— diagram of deviations from, 215
— origin of, 206
Granitell, 211
Granitic type, deviations from, 213
Granitite, 210
Granular diabase, 247
Granulite, 211, 295
Green, A. H., 37
Greenstone, 241
— tuffs, 249
Greisen, 211
Grinding of rock sections, 65
Grit, 17, 19
Groth, 140, 150
Groundmass, term defined, 168
Gas inclosures, 164
Gümbel, 155, 248
Gypsum, 304

HALB-GRANIT, 211
Hälleflinta, 209, 214
Halifax flags, 278
Hammers, 40
Hassock, 281
Hastings sands, 280
Haplite, 211

INT

Hardness, scale of, 44
Hart Hill, sandstone of, 278
Hartley, W. N., 164
Haselgebirge, 300
Haughton, S., 113, 207, 208, 257
Hauyne, 112
— and nosean, 232
— basalt, 257
Hauynophyr, 257
Hawaii, filiform lava of, 186
— obsidian of, 191
Headon Hill sands, 281
Hematite, 155, 304
Henslow, 212
Hertfordshire, puddingstone of, 301
Heulandite, 160
Hexagonal system, cleavages in, 172
Hills and mountains, causes affecting their forms, 29
— and valleys, formation of, 28
Himalayas, sandstone of, 281
Homogeneous vitreous rocks, 180
Horizontal strata, map and section of district composed of, 22
— — denudation of, 23
Hornblende, 128
— andesite, 234
— schist, 295, 298
— syenite, 217
Hornblendic granite, 203
Hydrometamorphism, 208
Hypersthene, 119
— andesite, 234
Hypersthenite, 250
Hull, E., 257, 277
Hungary, tridymite of, 152
Hunt, R., 280
— T. Sterry, 36, 271
— — — theory of felspars, 95

ICE, 304
Idocrase, 142
Inclined strata, denudation of, 23
Inclosures in quartz, 150
— of fluid, 164
— of glass, 165
Indicator for eye-piece, 52
Internal heat of the earth, evidence of, 10
Interpositions in felspars, 95

INT

Intrusive sheets distinguished from lava-flows, 32
Iron glance, 155
— ores, 304
— pyrites, 156
Isle of Wight, tertiary limestone of, 289
Isotropic substances, 76
Itacolumite, 297
Italian marbles, 287, 293

JADE, 131
Jasper, porcelain, 290
Jervis, W. P., 293
Jointing, 12
Jordan's section cutter, 61
Judd, J. W., 37, 191, 193, 262, 280
Jukes, J. B., 13
Jurassic sandstones, 280

KALKOWSKY, 171
Kaolin, 203, 272, 284
— formation of, 30
Kaolinite, 284
Katzenbuckel, nephelinite of, 255
Kengott, 120, 219
Kersantite, 220, 239
Kersanton, 220, 239
Ketton stone, 286
Keuper sandstone, 279
Kieselschiefer, 293
Kilanea, lavas of, 199
Kinahan, G. H., 209, 238
Kinzigite, 262
Kjerulf, 257
Klaproth, 228
Klein's quartz-plate, 58
Knaresborough, calcareous tufa of, 302
Knotenschiefer, 292
Kobell, V., stauroscope of, 81
Kupferberg, bronzite of, 121
Kyanite, 144
— in quartz, 151

LAACHER See, hauyne-basalt of, 257
— — hauyne of, 114
Labelling specimens, 42
Labradorite, 99
Lake District, mica traps of, 220
— — volcanic ejectamenta of, 268

MAR

Laminar fission, 12
Lapilli, 267
Lasaulx, A. von, 59, 103, 114, 119, 133, 195, 216, 218, 231, 232, 234, 247, 255, 264, 290
Laterite, 272
Lava-flows distinguished from intrusive sheets, 32
Lavas of Etna, 255
— — Vesuvius, 256
Lebour, G. A., 32
Leeson's microscope, 59
Lepidolite, 134
Lepidomelane, 136
Leptinite, 211, 295
Leucilite, 256
Leucite, 108
— basalt, 256
Leucitophyr, 256
Leucoxene, 155
Levigation, 73
Lévy, Michel, 152
Lherzolite, 263
Liassic clays, 284
— limestones, 286
Limestones, 17, 20
Limestone, carboniferous, 285
— cretaceous, 288
— crystalline, 287, 293
— Devonian, 285
— magnesian, 20, 285
— oolitic, 286
— Silurian, 285
— tertiary, 289
Limonite, 156
Lincolnshire oolites, 286
Lipari, obsidian of, 183, 189
Liparite, 222
Lithia mica, 134
Lithoidite, 222
Llanberis quarries, 283
London clay, 284
Luxullianite, 210
Lydian stone, 293
Lyell, Sir C., 11, 37, 303, 305

MACHINES for grinding microscopic sections, 61
Macle, 143
Magma, term defined, 168
Magnesian limestone, 20, 285
Magnefite, 153
Magnets, 45
Maltwood's finder, 51
Marble, encrinital, 285
Marbles, Italian, 287, 293

NEP

Marbles, microscopic character of, 287
Marcasite, 157
Margarodite, 135
Marialite, 112
Marine denudation, effects of, 25
Marl, 19
Maskelyne, N. S., 120
Matlock, calcareous tufa of, 302
Medway, mud of, 285
Meionite, 112
Melaphyre, 260
Merrifield, C.W., 259
Metamorphism, 36, 208
Mica basalt, 257
Micaceous felstone, 209
Micas, 132
Mica schist, 297
— syenite, 218
Mica Trap, 220
Microcline, 96   [170
Micro-crystalline matter,
— felsitic matter, 171
Microliths, 162, 185
— in perlite, 194
Microscopes, 46
Microscopic analysis of E. Boricky, 100
— preparations, 59
Millstone lavas, 256
Mineral deposits, 303
Minerals, optical characters of, 74
— rock-forming, 86
Minette, 219
Miocene sandstones, 281
Missouri, hematite of, 156, 304
Möhl, H., 152, 252
Monoclinic system, cleavages in, 172
Moraines, 301
Moroxite, 147
Morris, J., 281
Mountains and hills, causes affecting their forms, 29
Mounting sections of rocks, &c., 69
Mudstones, 17
Murchison, Sir R. I., 212
Muscovite, 133
Museums, arrangement of rock collections in, 43

NAPLES, piperno of, 233
Nassau, schalstein of, 248
Natrolite, 159
Negative crystals, 165
Nepheline, 104
— alteration of, into natrolite, 159

Index. 319

NEP

Nepheline basalt, 255
Nephelinite, 255
Neurode, hypersthenite of, 251
Nevada, U.S., vitreous rocks of, 184
New red conglomerate, 301
—— sandstone, 279
Newton, E. T., 71
Niedermendig, lava of, 256
Nile mud, 282
Norite, 251
Normal sedimentary rocks, 275
Northampton sand, 280
North Elmsley, Canada, moroxite of, 147
Nosean, 112
— and hauyne, 232
Nose-pieces, 52

OBJECT-glasses, 48
Obsidian, 186
— crystallites in, 187
— spherulites in, 188
Old red conglomerate, 302
—— sandstone, 277
Oligoclase, 99
— diorite, 241
Olivine, 116
— alteration of, 264, 271
— basalt, 253
— gabbro, 249
Oolite, 286
Oolitic limestones, 286
Opacite, 166
Optical axes, 78
— properties of minerals, 74
———— determination of, 79
Orthoclase, 92
Orthoclastic felspars, 91
Osteocolla, 302

PACHUCHA, Mexico, tridymite from, 152
Palagonite rock, 272
— tuff, 272
Paragonite, 135
Paranthine, 111
Pearlite, 192
Pearlstone, 192
Pélé's hair, 186
Pelitic rocks, 299
Penck, A., 268
Pennant grit, 278
Penrhyn quarries, 283
Penrith sandstone, 279
Percy, J., 285
Perlite, 192

PYR

Perlitic structure, 182
—— in tachylyte, 194
Permian limestones, 285
—— sandstones, 278
Perthite, 97
Petrosilex, 209, 214
Petworth marble, 287
Phillips, J., 36
— J. A., 37, 70, 151, 211
Phlogopite, 135
Phonolite, 228
— classification of, 229
— conglomerate, 233
— tuff, 233
— wacké, 233
Phosphoric acid, detection of, 145
Picrite, 265
Pilot Knob, Missouri, 156
Piperno, 233
Pisolite, 18, 303
Pitchstone, 195
— devitrification of, 198
— felsitic, 197
— trachytic, 195
Plagioclase, 91
— basalt, 253
— enstatite rocks, 251
Plagioclastic felspars, 91
Pliny, 203, 217
Plutonic and volcanic rocks, 33
Pocket lens, 44
Polarisation of light, 75
Polarising apparatus, 48, 75
Poole clay, 284
Porcelain jasper, 290
Porphyrite, 237
Porphyritic structure in vitreous rocks, 185
Porphyroid, 296
Portland cement, 285
— oolites, 286
Potstone, 271
Poussin, Ch. de la Valée, 280
— and Rénard, 135
Practical value of petrological research, 3
Pre-Cambrian basalts, 262
Preliminary examination of rocks, 44
Propylite, 237
Protogine gneiss, 212, 295
— granite, 213
Provisional names applied to minerals, 166
Psammitic rocks, 299
Psephitic rocks, 299
Pseudo-chrysolite, 187
Pseudomorphs, 30
Puddingstone, 301
Pyrites, 156

ROU

Pyrometamorphism, 208
Pumice, 191, 267
Purbeck limestone, 286
— marble, 286
Pyramids, nummulitic limestone of the, 289
Pyroxene, 121

QUARTZ, 149
— diabase, 247
— inclosures in, 150
— porphyry, 210
— rhyolite, 223
— trachyte, 223
Quartzite, 293
Quartzose dacite, 234
Quartzless diabase, 245
— hornblende andesite, — trachyte, 223 [235

RAIN, effect of, upon rocks, 29
Rammelsberg, 150 [302
Ramsay, A. C., 36, 212,
Rath, G. vom, 108, 250
Reiss and Fritsch, 233
Rénard, A., 59, 86, 135, 142, 165, 280
Reusch, 149
Reyer, E., 38
Rhombic system, cleavages in, 172
Rhyolite, 178
Richthofen, Von, 178, 193, 237
Roberts, W. Chandler, 258
Rodwell, G. F., 267
Rosenbusch, H., 108, 142, 143, 144, 148, 152, 159, 160, 167, 169, 170, 171, 196, 199, 219, 221, 228, 232, 245, 249, 254, 264
Rosenbusch's microscope, 54
Roth, J., 178, 234
Rocche Rosse, obsidian of, 184, 189, 191
Rock salt, 304
—— in fluid inclosures, 165
Rochdale flags, 278
Roches moutonnées, 305
Rocks, condition under which formed, 6
— general characters of, 6
Rotheram stone, 278
Rothliegende, 278
Rothweil, analcime from, 160
Roto-mahana, 33, 303
Rounded crystals, 178
Rounding of stones, 16

Index.

Rutile, 147
— in quartz, 151

SAINT Bees sandstone, 279
Salt, crystals of, in fluid inclosures, 165
Sands, 17
Sand, volcanic, 267
Sandberger, 254
Sand-rock, 19, 280
Sandstones, 17, 276
— cambrian, 276
— carboniferous, 278
— cretaceous, 280
— Devonian, 277
— jurassic, 280
— old red, 277
— oolitic, 280
— Silurian, 276
— tertiary, 281
Sandwich Islands, lavas of, 199
Sanidine, 94
— rhyolite, 224
— trachyte, 224
Scapolite, 111
Scenery, on what dependent, 11
Schalstein, 248
Scheerer, 108
Schistose rocks, 293
Schmidt's goniometer, 53
Schorl, 138
— rock, 265, 298
— schist, 298
Scotch slates, 284
Scrope, G. Poulett, 37, 228
Sedimentary matter, sorting of, in water, 16
— rocks, 274
— — classification of, 19
— — defined, 15
Seifersdorf, minette of, 219
Seismology, 9
Semi-granite, 211
Serpentine 269
Sericite, 134
— schist, 296
Setton, les, 152
Shales, 17, 282
Sharp, D., 36
Silica, globular condition of, 152
— percentage of in eruptive rocks, 177
Siliceous breccia, 19
— limestone, 20
— sinter, 19, 33, 303
Silurian flags, 283
— limestones, 285
— sandstones, 276

Silurian slates, 283
Sinter, siliceous, 19, 33, 303
Skiddaw slates, 283
Skye, hypersthenite of, 251
Slabs, 282
Slates, 17, 282
— Cambrian, 283
— Silurian, 283
Slaty cleavage, cause of, 35
Slicing rocks with diamond dust, 63
Slievenalargy, tachylyte of, 200
Smith, Lawrence, microscope of, 59
Sphene, 140
Sodalite, 112
— of Somma, 116
Sorby, H. C., 2, 36, 107, 142, 151, 224
Sorne, Isle of Mull, tachylyte of, 202
South Burgess, Canada, moroxite of, 147
Specimens, collection of, 39
— dressing of, 41
Specular iron, 155
Spherulites in obsidian, 188
Spherulitic structure, 183
Spilite, 247
Stache, 237
Statuary marble, 287
Staurolite slate, 292
Stauroscope, 81
Stauroscopic examination, 84
Stelzner, 171
Stockwerksporphyr, 211
Stone inclosures, 165
Strata, flexure of, 21
Stratigraphical breaks, 31
Streak of minerals, 45
Striations in labradorite, 102
Strike, term defined, 21
Structures in basalt, &c., due to contraction, 14, 259
— — vitreous rocks, 180
Structural planes, 12
Subsidence and elevation of land, 11
Syenite, 203, 217
Syenitic gneiss, 295
— granite, 203
Synclinal, term defined, 22
Szabo, 100

TABLE of cleavages in minerals, 171
Tachylyte, 199

Talc, 137
— schist, 298
Taupo Lake, siliceous sinter of, 303
Tawney, E. B., 262
Tephrite, 253
Tertiary limestones, 289
— sandstones, 281
Tetragonal system, cleavages in, 171
Thames, mud of, 285
Thermal springs, 33
Thickness of beds, measurement of, 24
— of the earth's crust, 10
Tin stone, 148
Tintagel Quarries, 284
Titaniferous iron, 154
Titanite, 140
Tivoli, travertine of, 303
Tolcsva, obsidian of, 181
Topaz, 142
Topley, W., 32, 281, 289
Tourmaline, 137
Trachy-phonolite, 279
Trachyte, 221
— proper, 225
Trachytes, classification of, 222
Trachytic pitchstone, 195
Travertine, 18, 303
Tremolite, 131
Triassic sandstones, 279
Trichites, 162, 185
Triclinic system, cleavages in, 172
Tridymite, 152
Tripoli, 281
Tschermak, 88, 95, 103
Tufa, calcareous, 18, 302
Tuff, greenstone, 249
— phonolite, 233
Twinning of calcspar, 149
— — felspars, 87, 102

UNCONFORMITY, stratigraphical, 31
United States, propylite of, 237
— — rhyolite of, 185

VARIOLITE, 248
Vibration, principal directions of, in crystals, 58
Vitreous rocks, 177
— — phenomena of fusion in, 179
Vélain, M., 152
Veltlin, hypersthenite of, 251
Verde, antique porphyry, 240

VES

Vesuvian, 142
— lavas, 257
Vesuvius, leucitophyrs of, 256
— nepheline of, 107
Viridite, 166
Vitreous eruptive rocks, 177
Vogelsang, H., 113, 119, 142, 163, 171, 197, 201
Volcanic phenomena, 38
— and plutonic rocks, defined, 33
— ashes, 267
— bombs, 267
— ejectamenta, 266
— sand, 267
Volcanoes, general characters of, 37
Vom Rath, 152
Vulcanicity, 9

WOL

WADSWORTH, U.S., rhyolite of, 185
Ward, J. C., 34, 144, 249, 291
Washing of fine deposits, 73
Watcombe clay, 284
Wealden marbles, 289
Weathering of rocks, 29
Weiss-stein, 211
Welsh slates, 283
Wernerite, 111
Whin Sill, 258
Wickersley stone, 278
Witham, H., sections of fossils first prepared by, 2
Wolf rock, 107
— — tridymite in, 152
Wolff, Th., 171, 266

ZWI

Wollaston's prism, 49
Woodward, H. B., 282
Woolwich and Reading clays, 284

YORKSHIRE flags, 278
— jurassic sandstones of, 280

ZEOLITES, 158
Zircon, 143
Zirkel, F., 106, 107, 108, 114, 121, 125, 129, 130, 146, 147, 160, 166, 167, 168, 184, 197, 201, 219, 221, 228, 237, 249, 251
Zwitter rock, 211

www.ingramcontent.com/pod-product-compliance
Lightning Source LLC
Chambersburg PA
CBHW021159230426
43667CB00006B/465